How to Gain Gain

Burkhard Vogel

How to Gain Gain

A Reference Book on Triodes in Audio
Pre-Amps

Springer

Dipl. -Ing. Burkhard Vogel
BUVOCON GmbH
70180 Stuttgart
Germany

ISBN: 978-3-642-08904-6   e-ISBN: 978-3-540-69505-9

Cover design: WMXDesign GmbH

Printed on acid-free paper

9 8 7 6 5 4 3 2 1

springer.com

To all valve enthusiasts

"A valve and a piece of wire - and the whole stuff will play!"

My friend Otto, 1971 -
after a valve amp crash

"Math is for solid-state; poetry is for tube audio"[1]

Tube Cad Journal, Nov. 2004

"God doesn't play at dice"

Albert Einstein

---

[1] Worth reading: "Science versus subjectivism in audio engineering", Douglas Self,
Electronics & Wireless World, July 1988, p. 692 ff

Preface

The today's (2008) audio (or low-frequency) amplifier world is nearly 100% "contaminated" by silicon solid-state components that ensure proper amplification for any signal from any source (CD, DVD, Vinyl, MP3, Radio, TV, etc.) to ones ears. A broad range of design supporting literature[2] and other information sources and tools like eg. the internet or component producer's applications help design engineers world-wide to come up with the right products at the right time on the right markets - hopefully.

The percentage difference between 100% and nearly 100% is filled with gain making elements that are far away from silicon: vacuum valves and audio transformers. It's an interesting fact that still many CD and/or DVD producing studios work with valve and transformer driven equipment like eg. the compressor types Teletronik LA-2A or UREI 1176LN or they bring into action valve driven microphones (eg. Neumann U47 or Rode Classic II) and mixers at the front-end of the recording chain. Not to forget all the valve powered measurement equipment (from eg. Brüel & Kjær, Tektronics, etc.) still in use.

Because of their outstanding sound quality (eventually mainly caused by an even harmonics distortion effect) most of the involved valves in pre-amp stages are triodes or pentodes configured as triodes. I won't debate whether silicon or vacuum sounds better - this would be purely subjective. But the world-wide growing sales revenues for vacuum based sound and reproduction equipment is an astonishing thing for me and gives a sound related answer by purse or credit card. And another thing surprises me as well: despite the totally different analogue world there are enough young engineers (and senior ones - of course) of our totally digital era that are willing to struggle with such an old fashioned technology by creating superb sounding electronic instruments.

Although it might look as if - but it has nothing to do with Black Art! It's simply the transfer of a specific know-how into life by enthusiasts. A know-how that seems to be no longer part of training courses of universities and colleges, that is threatened to get lost if we don't work hard to stop this evolution by bringing out ready and easy to use modern literature and software tools.

Therefore, the following chapters offer formulae to calculate certain building blocks of valve amplifiers - but for pre-amp purposes only. In nearly all cases detailed derivations are also given. All what's needed are the data sheet figures of the triode's $(_t)$ mutual conductance $g_{m.t}$, the gain factor μ_t and the internal plate resistance $r_{a.t}$. To calculate frequency and phase responses of gain stages the different data sheet presented input and output capacitances have to be taken into account as well.

---

[2]   Inter alia: "Electronic Circuits", U. Tietze, C. Schenk, 2nd edition, Springer 2008, ISBN 978-3-540-00429-5

It must be pointed out that all formulae are based on certain assumptions. The most important one is the one that defines the DC biasing conditions of these active devices. The conditions were assumed to be those of the A-class operating point settings: the plate DC current never switches off for positive and negative parts of the input signal. In other words: the A-class operating point is located in the (~middle of the) most linear part of the V_g / I_a versus plate voltage V_a diagram of the respective valve. This is the only way to guarantee that the triode's data sheet figures for mutual conductance $g_{m.t}$, gain μ_t and internal plate resistance $r_{a.t}$ can be taken as so-called constants (always valid: $\mu_t = g_{m.t}$ x $r_{a.t}$) and can be used for our calculation purposes.

Other biasing classes (B = total plate/cathode current switch-off for negative signal parts or AB = a tiny quiescent current is allowed to flow through plate and cathode for negative signal parts) or the use of other operating points on the V_g / I_a versus plate voltage V_a characteristic plot need certain additional measurements or graphical approaches to get the right values for the above shown valve constants. Having gone through these processes the newly generated $g_{m.t.new}$, $\mu_{t.new}$[3] and $r_{a.t.new}$ figures should be used for further calculation purposes. The given formulae won't change and will look the same.

I do not dive into the valve's DC biasing mechanics because they can easily be studied with the help of a broad range of literature[4]. But in that range of books and magazines[5] I miss a summary of all the gain producing possibilities of triodes on one spot. That will be the only matter of the following pages. The respective formulae were derived from equivalent circuits by application of Ohm's[6] and Kirchhof's[7] laws. It will be demonstrated in detail in Chapters 1... 4. These approaches lead to certain amp building blocks around one valve. The formulae for gain stages that incorporate more than one valve (eg. SRPP or cascoded gain stages, etc.) will mostly be derived from these building blocks.

In addition, MathCad[8] (MCD) worksheets as a part of each chapter allow easy follow-up and application of the respective formulae for any kind of triode. The calculations show results with 3 digits after the decimal point. The only reason for that is to demonstrate - from time to time - (tiny) differences with other calculation results. In reality, even a calculation result of one digit after the decimal point wouldn't present the whole truth because the tolerances of valves are a magnitude away from precision. But this fact didn't - nor doesn't it today - prevent engineers from designing extremely precise working analogue amps and other electronic valve driven devices. That's why, on the other hand, the calculation approaches offered are not far away from reality. For fast estimations many rules of thumbs are offered too.

I'm sure I only did treat a limited selection of possible building blocks for triode driven amps. That's why all readers are invited to not hesitate to send to the editors their know-how on additional triode amp stage solutions - including the mathematical derivations that are needed to understand how they work (à la the presented MathCad worksheets). This book should become the collection of everything what's of interest on this specific design field. The next edition will come out with these additional designs.

---

[3]   in contrast to respective "on dit(s)" a change of the bias point also means a change of μ_t; it is
         not a constant at all (see also Chapter 16 plots!)
[4]   Inter alia: "Valve amplifiers", Morgan Jones, Newnes, ISBN 0-7506-2337-3
[5]   Inter alia: "Tube Cad Journal"
[6]   a) 1Ω *1A = 1V or - generally spoken - R*I = V
[7]   a) The sum of all currents in a circuit's node equals zero;
         b) The sum of all voltages (= potential differences) in a circuit's closed loop equals zero.
[8]   MathCad is a registered trademark of MathSoft Engineering & Education Inc.,
         since 2006 part of Parametrics Technology Corporation (PTC), Ma., USA

To sum-up the aims of this book:

➤ building up a collection of triode amp stage alternatives with satisfactory mathematical demonstration on how they work via derivations and transfer functions,

➤ to make things less complex the transfer functions are derived from rather simplified equivalent circuits, thus, saving a lot of energy by paying for it with tiny frequency and phase response errors, especially at the ends of the audio band,

➤ it's always better to calculate first - and spent money for expensive components later - instead of playing around with dice-type trial and error.

Contents

Chapter 1 The Common Cathode Gain Stage (CCS)

1.1 Circuit diagram

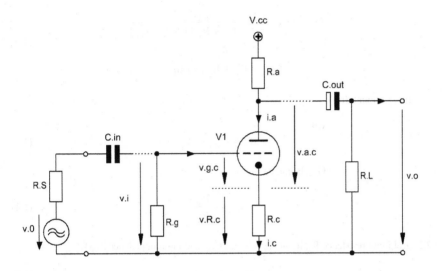

Figure 1.1 Basic design of the Common Cathode Gain Stage (CCS)

1.2   Basic formulae (excl. stage load (R_L)-effect)

1.2.1 gain G_t:

$$G_t = -\frac{v_o}{v_i} \tag{1.1}$$

in terms of $g_{m.t}$ and μ_t G_t becomes:

$$
\begin{aligned}
G_t &= -g_{m.t} \frac{r_{a.t} R_a}{r_{a.t} + R_a + R_c \left(1 + g_{m.t}\, r_{a.t}\right)} \\
&= -\mu_t \frac{R_a}{r_{a.t} + R_a + R_c \left(1 + \mu_t\right)}
\end{aligned}
\tag{1.2}
$$

rule of thumb (rot):

with

$$\mu_t\, R_c \gg r_{a.t} + R_a \tag{1.3}$$

$G_{t.rot}$ becomes:

$$G_{t.rot} = -\frac{R_a}{R_c} \tag{1.4}$$

1.2.2 grid input resistance R_g, input capacitance $C_{i.tot}$ and input impedance $Z_{i.g}(f)$:

$$Z_{i.g}(f) = R_g \parallel C_{i.tot} \tag{1.5}$$

$$C_{i.tot} = \left(1 + |G_t|\right) C_M + C_{g.c} + C_{stray} \tag{1.6}$$

1.2.3 plate output resistance $R_{o.a}$ and output impedance $Z_{o.a}(f)$ (by ignoring tiny R_c-effects):

$$R_{o.a} = R_a \parallel r_{o.a.t} \tag{1.7}$$

$$r_{o.a.t} = r_{a.t} + \left(1 + \mu_t\right) R_c \tag{1.8}$$

$$Z_{o.a}(f) = R_{o.a} \parallel C_{o.tot} \tag{1.9}$$

1.2.4 cathode output resistance $R_{o.c}$:

$$R_{o.c} = r_{c.t} \parallel R_c \tag{1.10}$$

$$r_{c.t} = \frac{R_a + r_{a.t}}{\mu_t + 1} \tag{1.11}$$

1.3   Derivations

1.3.1 Equivalent circuit for derivation purposes

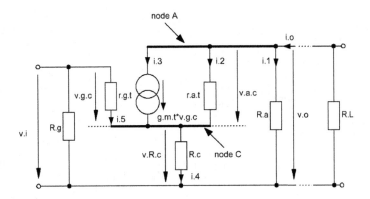

Figure 1.2 Equivalent circuit of Figure 1.1

1.3.2 Derivation

1.3.2.1   gain G_t:

sum of currents at node A:

$$i_o = i_1 + i_2 + i_3 \tag{1.12}$$

without R_L we can set $i_o = 0$, hence

$$-i_1 = i_2 + i_3 \tag{1.13}$$

$$i_1 = \frac{v_o}{R_a} \tag{1.14}$$

$$i_2 = \frac{v_{a.c}}{r_{a.t}} = \frac{v_o - v_{R.c}}{r_{a.t}} \tag{1.15}$$

$$i_3 = g_{m.t}\, v_{g.c} = g_{m.t}\left(v_i - v_{R.c}\right) \tag{1.16}$$

sum of currents at node C:

$$i_4 = i_2 + i_3 + i_5 \tag{1.17}$$

$r_{g.t}$ can be assumed as infinite, hence we can set $i_5 = 0$, thus

$$i_4 = \frac{v_{R.c}}{R_c} = \frac{v_o - v_{R.c}}{r_{a.t}} + g_{m.t}\left(v_i - v_{R.c}\right) \tag{1.18}$$

rearrangement leads to:

$$i_4 = -i_1 \tag{1.19}$$

$$G_t = -\frac{v_o}{v_i} \tag{1.20}$$

$$g_{m.t}\left(v_i + v_o\frac{R_c}{R_a}\right) + \frac{v_o + v_o\dfrac{R_c}{R_a}}{r_{a.t}} = -\frac{v_o}{R_a} \tag{1.21}$$

$$g_{m.t}v_i = -v_o\left(\frac{1}{R_a} + g_{m.t}\frac{R_c}{R_a} + \frac{1 + \dfrac{R_c}{R_a}}{r_{a.t}}\right) \tag{1.22}$$

$$\Rightarrow G_t = -g_{m.t}\frac{r_{a.t}\,R_a}{r_{a.t} + R_a + R_c\left(1 + g_{m.t}\,r_{a.t}\right)} \tag{1.23}$$

with the general triode equation G_t in terms of μ_t becomes:

$$g_{m.t}\,r_{a.t} = \mu_t \tag{1.24}$$

$$\Rightarrow G_t = -\mu_t\frac{R_a}{r_{a.t} + R_a + R_c\left(1 + \mu_t\right)} \tag{1.25}$$

$R_{a.eff}$ replaces R_a in all calculations that include R_L as gain stage output load!

$$R_{a.eff} = R_a \parallel R_L \tag{1.26}$$

$$G_{t.eff} = -g_{m.t}\frac{r_{a.t}\,R_{a.eff}}{r_{a.t} + R_{a.eff} + R_c\left(1 + g_{m.t}\,r_{a.t}\right)} \tag{1.27}$$

$$G_{t.eff} = -\mu_t\frac{R_{a.eff}}{r_{a.t} + R_{a.eff} + R_c\left(1 + \mu_t\right)} \tag{1.28}$$

1.3.2.2   the grid input impedance $Z_{i.g}(f)$ becomes:

$$Z_{i.g}\left(f\right) = r_{g.t} \parallel R_g \parallel C_{i.tot} \tag{1.29}$$

$$i_5 = 0$$

$$\Rightarrow \quad r_{g.t} = \infty \tag{1.30}$$

$$\Rightarrow \quad Z_{i.g}(f) = R_g \parallel C_{i.tot}$$

1.3.2.3   the effective plate output resistance $R_{o.a.eff}$ and impedance $Z_{o.a.eff}(f)$ become:

$$i_4 = i_2 + i_3 = g_{m.t}\, v_{g.c} + \frac{v_{a.c}}{r_{a.t}}$$

$$i_4 = g_{m.t}\left(v_i - v_{R.c}\right) + \frac{v_o - v_{R.c}}{r_{a.t}} \tag{1.31}$$

setting $v_i = 0$ leads to:

$$r_{a.t}\, i_4 = v_o - \left(1 + \mu_t\right) v_{R.c}$$

$$r_{a.t} = \frac{v_o}{i_4} - \left(1 + \mu_t\right) R_c$$

$$r_{o.a.t} = \frac{v_o}{i_4} \tag{1.32}$$

$$= r_{a.t} + \left(1 + \mu_t\right) R_c$$

$$R_{o.a} = r_{o.a.t} \parallel R_a$$

$$\Rightarrow Z_{o.a}(f) = R_{o.a} \parallel C_o \tag{1.33}$$

$$R_{o.a.eff} = r_{o.a.t} \parallel R_{a.eff}$$

$$\Rightarrow Z_{o.a.eff}(f) = R_{o.a.eff} \parallel C_{o.tot} \tag{1.34}$$

1.3.2.4   the effective cathode output resistance $R_{o.c.eff}$ looks as follows[1]:

$$r_{c.t} = \frac{r_{a.t} + R_a}{1 + \mu_t} \tag{1.35}$$

$$R_{o.c} = r_{c.t} \parallel R_c \tag{1.36}$$

$$\Rightarrow r_{c.t.eff} = \frac{r_{a.t} + R_{a.eff}}{1 + \mu_t} \tag{1.37}$$

$$\Rightarrow R_{o.c.eff} = r_{c.t.eff} \parallel R_c \tag{1.38}$$

---

[1]   detailed derivation see Chapter 4 "The common grid stage (CGS)"

1.3.2.5   total input capacitance $C_{i.tot}$, total output capacitance $C_{o.tot}$ and Miller
capacitance C_M:

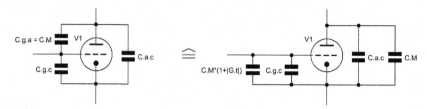

Figure 1.3 Valve capacitances

Data sheet figures:

$C_{g.c}$ = C between grid and cathode (rather often expanded by
additional heater and shield capacitances for the CGS gain
stage situation - see respective chapter)

$C_{g.a}$ = C between grid and plate

$C_{a.c}$ = C between plate and cathode (rather often expanded by
additional heater and shield capacitances for the CGS gain
stage situation - see respective chapter)

Calculation relevant figures:

C_M = $C_{g.a}$ = Miller capacitance

C_{stray} = sum of several different capacitances (to be guessed)[2]

$C_{i.tot}$ = total input capacitance

$C_{o.tot}$ = total output capacitance

$$C_{i.tot} = \left(1 + |G_t|\right) C_{g.a} \parallel C_{g.c} \parallel C_{stray} \qquad (1.39)$$

$$C_{o.tot} = C_{a.c} \parallel C_M \qquad (1.40)$$

---

[2]   Guessed input sum of stray capacitances plus all other existing valve capacitances that
were not specifically mentioned in the calculation course, eg. capacitances from grid or
plate or cathode to heater or screen or to both or to the equivalent points of a second
system of a double triode

1.4   Gain stage frequency and phase response calculations

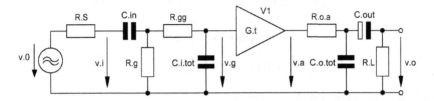

Figure 1.4 Simplified[3] equivalent circuit of Figure 1.1 - including all
frequency and phase response relevant components

1.4.1 gain stage input transfer function $T_i(f)$ and phase $\varphi_i(f)$ - including source
resistance R_S and an oscillation preventing resistor $R_{gg} \ll R_g$:

$$T_i(f) = \frac{v_g}{v_0} \tag{1.41}$$

$$\varphi_i(f) = \arctan\left\{\frac{\operatorname{Im}\left[T_i(f)\right]}{\operatorname{Re}\left[T_i(f)\right]}\right\} \tag{1.42}$$

$$T_i(f) = \frac{Z2(f)\left(\dfrac{1}{R_g} + \dfrac{1}{R_{gg} + Z2(f)}\right)^{-1}}{\left(R_{gg} + Z2(f)\right)\left[R_S + Z1(f) + \left(\dfrac{1}{R_g} + \dfrac{1}{R_{gg} + Z2(f)}\right)^{-1}\right]} \tag{1.43}$$

$$\begin{aligned}
Z1(f) &= \left(2j\pi f\, C_{in}\right)^{-1} \\
Z2(f) &= \left(2j\pi f\, C_{i.tot}\right)^{-1}
\end{aligned} \tag{1.44}$$

1.4.2 gain stage output transfer function $T_o(f)$ and phase $\varphi_o(f)$:

$$T_o(f) = \frac{v_o}{v_a} \tag{1.45}$$

$$\varphi_o(f) = \arctan\left\{\frac{\operatorname{Im}\left[T_o(f)\right]}{\operatorname{Re}\left[T_o(f)\right]}\right\} \tag{1.46}$$

---

[3]   "Simplified" because of footnote 2 of this chapter and the fact that the insertion of C_o
the way I do it "ignores" the existence of R_c. But this makes the calculation less complex
and will also lead to results not far away from the audio band reality.

$$T_o(f) = \left(\frac{Z3(f) \parallel (Z4(f) + R_L)}{R_{o.a} + \left[Z3(f) \parallel (Z4(f) + R_L) \right]} \right) \left(\frac{R_L}{R_L + Z4(f)} \right) \qquad (1.47)$$

$$Z3(f) = (2j\pi f\, C_{o.tot})^{-1}$$
$$Z4(f) = (2j\pi f\, C_{out})^{-1} \qquad (1.48)$$

1.4.3 fundamental gain stage phase shift $\varphi_{G.t}(f)$:

$$\varphi_{G.t}(f) = -180° \qquad (1.49)$$

1.4.4 gain stage transfer function $T_{tot}(f)$ and phase $\varphi_{tot}(f)$:

$$T_{tot}(f) = T_i(f)\, T_o(f)\, G_t \qquad (1.50)$$

$$\varphi_{tot}(f) = \varphi_i(f) + \varphi_o(f) + \varphi_{G.t}(f) \qquad (1.51)$$

1.5   Example with ECC83 / 12AX7 (83):

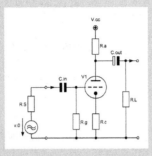

Figure 1.5 CCS example circuitry

1.5.1   Triode bias data:

$I_{a.83} = 1.2 \cdot 10^{-3} A$ $V_{cc.83} = 370 V$ $V_{a.83} = 250 V$ $V_{g.83} = -2V$

1.5.2   Triode valve constants:

$g_{m.83} := 1.6 \cdot 10^{-3} \cdot S$ $\mu_{83} := 100$ $r_{a.83} := 62.5 \cdot 10^{3} \Omega$

$C_{g.c.83} := 1.65 \cdot 10^{-12} F$ $C_{g.a.83} := 1.6 \cdot 10^{-12} F$ $C_{a.c.83} := 0.33 \cdot 10^{-12} F$

1.5.3   Circuit variables:

$R_{a.83} := 100 \cdot 10^{3} \Omega$ $R_{c.83} := 1.6 \cdot 10^{3} \Omega$ $R_{g.83} := 1 \cdot 10^{6} \Omega$

$R_{S} := 1 \cdot 10^{3} \Omega$ $R_{L} := 1 \cdot 10^{6} \Omega$ $R_{gg.83} := 2.2 \cdot 10^{3} \Omega$

$C_{stray.83} := 10 \cdot 10^{-12} F$ $C_{in.83} := 100 \cdot 10^{-9} F$ $C_{out.83} := 100 \cdot 10^{-9} F$

1.5.4   Calculation relevant data:

frequency range f for the below shown graphs: $f := 10Hz, 20Hz.. 20000Hz$

$h := 1000Hz$

1.5.5   Gain G_t and effective gain $G_{t.eff}$:

$$G_{83} := -g_{m.83} \frac{R_{a.83} \cdot r_{a.83}}{R_{a.83} + r_{a.83} + \left(1 + g_{m.83} \cdot r_{a.83}\right) \cdot R_{c.83}}$$

$G_{83} = -30.855 \times 10^{0}$

$$G_{83.e} := 20 \cdot \log\left(\left|G_{83}\right|\right)$$

$G_{83.e} = 29.786 \times 10^{0}$ [dB]

$$R_{a.83.eff} := \left(\frac{1}{R_{a.83}} + \frac{1}{R_L} \right)^{-1}$$

$$R_{a.83.eff} = 90.909 \times 10^3 \, \Omega$$

$$G_{83.eff} := -\mu_{83} \cdot \frac{R_{a.83.eff}}{R_{a.83.eff} + r_{a.83} + (1 + \mu_{83}) \cdot R_{c.83}}$$

$$G_{83.eff} = -28.859 \times 10^0$$

$$G_{83.eff.e} := 20 \cdot \log \left(\left| G_{83.eff} \right| \right)$$

$$G_{83.eff.e} = 29.206 \times 10^0 \, [dB]$$

1.5.6   Gain $G_{t.rot}$:

$$\mu_{83} \cdot R_{c.83} = 160 \times 10^3 \, \Omega$$

$$r_{a.83} + R_{a.83} = 162.5 \times 10^3 \, \Omega$$

$$G_{83.rot} := -\frac{R_{a.83}}{R_{c.83}}$$

$$G_{83.rot} = -62.5 \times 10^0$$

this result contradicts equ. (1.3), thus, a gain calculation via $G_{t.rot}$ makes no sense!

1.5.7   Specific resistances:

$$r_{o.a.83} := r_{a.83} + (\mu_{83} + 1) \cdot R_{c.83}$$

$$r_{o.a.83} = 224.1 \times 10^3 \, \Omega$$

$$R_{o.a.83} := \left[\left(\frac{1}{r_{o.a.83}} + \frac{1}{R_{a.83}} \right)^{-1} \right]$$

$$R_{o.a.83} = 69.145 \times 10^3 \, \Omega$$

$$R_{o.a.83.eff} := \left(\frac{1}{r_{o.a.83}} + \frac{1}{R_{a.83.eff}} \right)^{-1}$$

$$R_{o.a.83.eff} = 64.673 \times 10^3 \, \Omega$$

$$r_{c.83} := \frac{r_{a.83} + R_{a.83}}{1 + \mu_{83}}$$

$$r_{c.83} = 1.609 \times 10^3 \, \Omega$$

$$R_{o.c.83} := \left(\frac{1}{r_{c.83}} + \frac{1}{R_{c.83}} \right)^{-1}$$

$$R_{o.c.83} = 802.222 \times 10^0 \, \Omega$$

$$r_{c.83.eff} := \frac{r_{a.83} + R_{a.83.eff}}{1 + \mu_{83}}$$

$$r_{c.83.eff} = 1.519 \times 10^3 \, \Omega$$

$$R_{o.c.83.eff} := \left(\frac{1}{r_{c.83.eff}} + \frac{1}{R_{c.83}} \right)^{-1}$$

$$R_{o.c.83.eff} = 779.198 \times 10^0 \, \Omega$$

1.5.8   Specific capacitances :

$$C_{i.tot.83} := \left(1 + \left| G_{83} \right| \right) \cdot C_{g.a.83} + C_{g.c.83} + C_{stray.83}$$

$$C_{i.tot.83} = 62.617 \times 10^{-12} \, F$$

$$C_{o.tot.83} := C_{a.c.83} + C_{g.a.83}$$

$$C_{o.tot.83} = 1.93 \times 10^{-12} \, F$$

➢   MCD Worksheet I: CCS calculations   Page 3

1.5.9   Gain stage frequency and phase response:

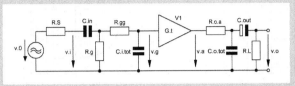

Figure 1.6 = Figure 1.4

$$Z1(f) := \frac{1}{2j \cdot \pi \cdot f \cdot C_{in.83}}$$

$$Z2(f) := \frac{1}{2j \cdot \pi \cdot f \cdot C_{i.tot.83}}$$

$$T_i(f) := \frac{Z2(f) \cdot \left(\dfrac{1}{R_{g.83}} + \dfrac{1}{R_{gg.83} + Z2(f)} \right)^{-1}}{\left(Z2(f) + R_{gg.83} \right) \left[R_S + Z1(f) + \left(\dfrac{1}{R_{g.83}} + \dfrac{1}{R_{gg.83} + Z2(f)} \right)^{-1} \right]}$$

$$\phi_i(f) := \operatorname{atan}\left(\frac{\operatorname{Im}\left(T_i(f) \right)}{\operatorname{Re}\left(T_i(f) \right)} \right)$$

$$T_{i.e}(f) := 20 \cdot \log\left(\left| T_i(f) \right| \right)$$

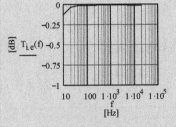

Figure 1.7 Transfer of i/p network

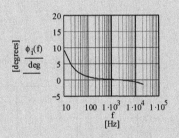

Figure 1.8 Phase of i/p network

$$Z3(f) := \frac{1}{2j \cdot \pi \cdot f \cdot C_{o.tot.83}}$$

$$Z4(f) := \frac{1}{2j \cdot \pi \cdot f \cdot C_{out.83}}$$

$$T_o(f) := \frac{\left(\dfrac{1}{Z3(f)} + \dfrac{1}{Z4(f) + R_L} \right)^{-1}}{R_{o.a.83} + \left(\dfrac{1}{Z3(f)} + \dfrac{1}{Z4(f) + R_L} \right)^{-1}} \cdot \frac{R_L}{R_L + Z4(f)}$$

$$\phi_o(f) := \operatorname{atan}\left(\frac{\operatorname{Im}\left(T_o(f) \right)}{\operatorname{Re}\left(T_o(f) \right)} \right)$$

$$T_{o.e}(f) := 20 \cdot \log\left(\left| T_o(f) \right| \right)$$

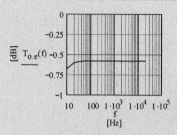

Figure 1.9 Transfer of o/p network   Figure 1.10 Phase of o/p network

$T_{tot.83}(f) := T_i(f) \cdot T_o(f) \cdot G_{83}$ $\phi_{G.83}(f) := -180 deg$

$T_{tot.83.e}(f) := 20 \cdot \log\left(\left|T_{tot.83}(f)\right|\right)$ $\phi_{tot.83}(f) := \phi_i(f) + \phi_o(f) + \phi_{G.83}(f)$

1.5.10 Frequency and phase response plots:

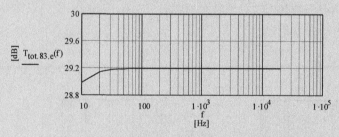

Figure 1.11 Frequency response of the whole CCS gain stage

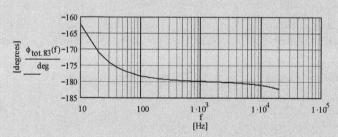

Figure 1.12 Phase response of the whole CCS gain stage

1.6   With changed circuitry components and valve constants the same calculation course
as above leads to an overall gain improvement of appr. 1.5dB (n = new):

$I_{a.83.n} = 0.5 \cdot 10^{-3} A$ $V_{cc.83.n} = 200 V$ $V_{a.83.n} = 100 V$ $V_{g.83.n} = -1V$

$g_{m.83.n} := 0.9 \cdot 10^{-3} \cdot S$ $\mu_{83.n} := 95$ $r_{a.83.n} := 105 \cdot 10^3 \Omega$

$R_{a.83.n} := 200 \cdot 10^3 \Omega$ $R_{c.83.n} := 2 \cdot 10^3 \Omega$

1.7   Additional remarks

1.7.1 Valve constants change with changing plate DC current I_a and changing plate-cathode DC voltage V_a. Many data sheets of modern valves offer specific charts[4] that demonstrates these changes:

> ➢   "constants vs. I_a" for a selection of different V_as

1.7.2 Plate currents lower than the one for the A-class biasing point produce higher plate resistances $r_{a.t}$, thus, valve noise will increase as well - and vice verca.

1.7.3 The oscillation prevention resistor R_{gg} (mostly to be found fixed between the valve's grid and grid resistance R_g) also plays a negative role when talking about noise, especially in amplifiers for tiny signals, such as signals from cartridges for vinyl records or from microphones.

1.7.4 For further information on noise in audio amplifiers in general and - specifically - a deeper view into the valve noise problems please consult the author's book

> ➢   "The Sound of Silence" (= TSOS)[5].

1.7.5 For C_{in} and C_{out} a change from 100nF to 1µF will drastically improve the frequency and phase response of the amp stage.

---

[4]   see example valve constants charts in Chapter 16
[5]   TSOS see Chapter 17, Section 17.1

2.1   Circuit diagram

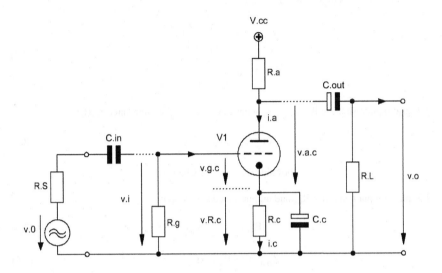

Figure 2.1 Basic design of the Common Cathode Gain Stage with
grounded Cathode via C_c (CCS+C_c)

2.2 Basic formulae (excl. stage load (R_L)-effect)

2.2.1 gain G_t (C_c acts as a complete short-circuit):

$$G_t = -\frac{v_o}{v_i} \tag{2.1}$$

gain G_t in terms of $g_{m.t}$ and μ_t becomes:

$$\begin{aligned} G_t &= -g_{m.t}\,\frac{r_{a.t}\,R_a}{r_{a.t}+R_a} \\ &= -\mu_t\,\frac{R_a}{r_{a.t}+R_a} \end{aligned} \tag{2.2}$$

2.2.2 grid input resistance R_g, input capacitance $C_{i.tot}$ and input impedance $Z_{i.g}(f)$:

$$Z_{i.g}(f) = R_g \parallel C_{i.tot} \tag{2.3}$$

formulae for $C_{i.tot}$: see Chapter 1

2.2.3 plate output resistance $R_{o.a}$ and output impedance $Z_{o.a}(f)$:

$$R_{o.a} = r_{a.t} \parallel R_a \tag{2.4}$$

$$Z_{o.a}(f) = R_{o.a} \parallel C_{o.tot} \tag{2.5}$$

formulae for C_o: see Chapter 1

2.2.4 cathode output resistance $R_{o.c}$:

$$\begin{aligned} R_{o.c} &= r_{c.t} \parallel R_c \\ r_{c.t} &= \frac{r_{a.t}+R_a}{1+\mu_t} \end{aligned} \tag{2.6}$$

2.2.5 capacitance C_c for a specific hp corner frequency $f_{c.opt}$:

$$C_c = \frac{1}{2\,\pi\,f_{c.opt}\,R_{o.c}} \tag{2.7}$$

2.3   Derivations

2.3.1 Equivalent circuit for derivation purposes:

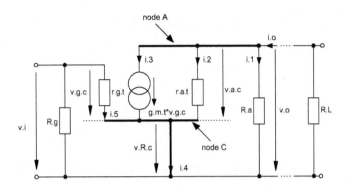

Figure 2.2 Equivalent circuit of Figure 2.1

2.3.2 Derivations:

With C_c of a size that acts as a short-circuit for $v_{R.c}$ in B_{20k} we can use all the respective formulae of Chapter 1 with $R_c = 0$, hence:

2.3.2.1   the gain G_t and effective gain $G_{t.eff}$ become:

$$G_t = -g_{m.t}\left(r_{a.t} \parallel R_a\right) \tag{2.8}$$

$$R_{a.eff} = R_a \parallel R_L \tag{2.9}$$

$$G_{t.eff} = -g_{m.t}\left(r_{a.t} \parallel R_{a.eff}\right) \tag{2.10}$$

2.3.2.2   the grid input impedance $Z_{i.g}(f)$ becomes:

$$Z_{i.g}\left(f\right) = \left|\frac{R_g}{1 + 2j\pi f\, R_g C_{i.tot}}\right| \tag{2.11}$$

2.3.2.3   the effective plate output resistance $R_{o.a.eff}$ and impedance $Z_{o.a.eff}(f)$ become:

$$R_{o.a} = r_{a.t} \parallel R_a$$
$$\Rightarrow\; Z_{o.a}\left(f\right) = R_{o.a} \parallel C_{o.tot} \tag{2.12}$$

$$R_{o.a.eff} = r_{a.t} \parallel R_{a.eff}$$
$$\Rightarrow\; Z_{o.a.eff}\left(f\right) = R_{o.a.eff} \parallel C_{o.tot} \tag{2.13}$$

$$Z_{o.a.eff}(f) = \left| \frac{R_{o.a.eff}}{1 + 2j\pi f\, R_{o.a.eff} C_o} \right| \tag{2.14}$$

2.3.2.4   cathode resistance $r_{c.t.eff}$ [1] and effective cathode output resistance $R_{o.c.eff}$ become:

$$r_{c.t.eff} = \frac{r_{a.t} + R_{a.eff}}{1 + \mu_t} \tag{2.15}$$

$$\begin{aligned} R_{o.c.eff} &= r_{c.t.eff} \parallel R_c \\ &= \frac{\dfrac{r_{a.t} + R_{a.eff}}{1 + \mu_t} R_c}{\dfrac{r_{a.t} + R_{a.eff}}{1 + \mu_t} + R_c} \end{aligned} \tag{2.16}$$

MCD symbolic evaluation "simplify" leads to

$$R_{o.c.eff} = \frac{(r_{a.t} + R_{a.eff}) R_c}{r_{a.t} + R_{a.eff} + (1 + \mu_t) R_c} \tag{2.17}$$

2.3.2.5   cathode capacitance C_c for a specific hp corner frequency f_c:

The audio band B_{20k} spreads from 20Hz to 20kHz. To ensure a flat frequency response in B_{20k} (± 0.1dB) as well as a phase response deviation of less than 1° at $f_c = 20$Hz the C_c calculation should be based on a corner frequency $f_{c.opt}$ that is a 100th of f_c.

$$f_c = 20\text{Hz}$$
$$f_{c.opt} = \frac{20\text{Hz}}{100} = 0.2\text{Hz} \tag{2.18}$$

$$C_c = \frac{1}{2\pi f_{c.opt} R_{o.c.eff}} \tag{2.19}$$

2.3.2.6   total input capacitance $C_{i.tot}$ and total output capacitance $C_{o.tot}$:

see respective equations and footnote[2] in Chapter 1 - 1.3.2.5

---

[1]   detailed derivation see Chapter 4 "The common grid stage (CGS)"
[2]   see footnote 2 in Chapter 1

2.4 Gain stage frequency and phase response calculations

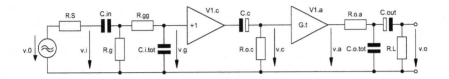

Figure 2.3 Simplified[3] equivalent circuit of Figure 2.1 - including all
frequency and phase response relevant components

2.4.1 gain stage input transfer function $T_i(f)$ and phase $\varphi_i(f)$ - including source resistance R_S and an oscillation preventing resistor $R_{gg} \ll R_g$:

see Chapter 1 - 1.4.1

2.4.2 gain stage output transfer function $T_o(f)$ and phase $\varphi_o(f)$:

see Chapter 1 - 1.4.2

2.4.3 gain stage cathode transfer function $T_c(f)$ and phase $\varphi_c(f)$:

$$T_c(f) = \left(\frac{R_{o.c.eff}}{R_{o.c.eff} + \dfrac{1}{2j\pi f\, C_c}} \right) \tag{2.20}$$

$$\varphi_c(f) = \arctan\left\{ \frac{\text{Im}\left[T_c(f)\right]}{\text{Re}\left[T_c(f)\right]} \right\} \tag{2.21}$$

2.4.4 fundamental gain stage phase shift $\varphi_{G.t}(f)$:

$$\varphi_{G.t}(f) = -180° \tag{2.22}$$

2.4.5 gain stage transfer function $T_{tot}(f)$ and phase $\varphi_{tot}(f)$:

$$T_{tot}(f) = T_i(f)\, T_o(f)\, T_c(f)\, G_t \tag{2.23}$$

$$\varphi_{tot}(f) = \varphi_i(f) + \varphi_o(f) + \varphi_c(f) + \varphi_{G.t}(f) \tag{2.24}$$

---

[3]   "Simplified" because of footnote 2 of Chapter 1. In addition: The simple C_c and $R_{o.c}$ network after the +1 amp $V1_c$ works correct the shown way only in the frequency range of $f_{c.opt}$ minus one octave. In lower frequency ranges than that the gain of the whole gain stage will reach the gain of a CCS gain stage with the same components but without C_c. But this frequency range lies far outside B_{20k}.

2.5   Example with ECC83 / 12AX7 (83):

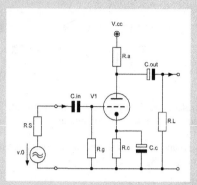

Figure 2.4 CCS+Cc example circuitry

2.5.1   Triode bias data:

$I_{a.83} = 1.2 \cdot 10^{-3} A$ $V_{cc.83} = 370 V$ $V_{a.83} = 250 V$ $V_{g.83} = -2 V$

2.5.2   Triode valve constants:

$g_{m.83} := 1.6 \cdot 10^{-3} \cdot S$ $\mu_{83} := 100$ $r_{a.83} := 62.5 \cdot 10^{3} \Omega$

$C_{g.c.83} := 1.65 \cdot 10^{-12} F$ $C_{g.a.83} := 1.6 \cdot 10^{-12} F$ $C_{a.c.83} := 0.33 \cdot 10^{-12} F$

$C_{o.tot.83} := C_{a.c.83} + C_{g.a.83}$ $C_{o.tot.83} = 1.93 \times 10^{-12} F$

2.5.3   Circuit variables:

$R_{a.83} := 100 \cdot 10^{3} \Omega$ $R_{c.83} := 1.6 \cdot 10^{3} \Omega$ $R_{g.83} := 1 \cdot 10^{6} \Omega$

$R_{S} := 1 \cdot 10^{3} \Omega$ $R_{L} := 1 \cdot 10^{6} \Omega$ $R_{gg.83} := 2.2 \cdot 10^{3} \Omega$

$C_{stray.83} := 10 \cdot 10^{-12} F$ $C_{in.83} := 100 \cdot 10^{-9} F$ $C_{out.83} := 100 \cdot 10^{-9} F$

2.5.4   Calculation relevant data:

frequency range f for the below shown graphs: $f := 10 Hz, 20 Hz .. 20000 Hz$

$h := 1000 Hz$

2.5.5   Gain G_t and effective gain $G_{t.eff}$:

$G_{83} := -g_{m.83} \cdot \dfrac{R_{a.83} \cdot r_{a.83}}{R_{a.83} + r_{a.83}}$ $G_{83} = -61.538 \times 10^{0}$

> ➢ MCD Worksheet II　　　　CCS+Cc calculations　　　　Page 2

$G_{83.e} := 20 \cdot \log\left(\left|G_{83}\right|\right)$ 　　　　　　$G_{83.e} = 35.783 \times 10^0$ [dB]

$R_{a.83.eff} := \left(\dfrac{1}{R_{a.83}} + \dfrac{1}{R_L}\right)^{-1}$ 　　　　$R_{a.83.eff} = 90.909 \times 10^3 \, \Omega$

$G_{83.eff} := -\mu_{83} \cdot \dfrac{R_{a.83.eff}}{R_{a.83.eff} + r_{a.83}}$ 　　　$G_{83.eff} = -59.259 \times 10^0$

$G_{83.eff.e} := 20 \cdot \log\left(\left|G_{83.eff}\right|\right)$ 　　　　$G_{83.eff.e} = 35.455 \times 10^0$ [dB]

2.5.6　Specific resistances:

$R_{o.a.83} := \left[\left(\dfrac{1}{r_{a.83}} + \dfrac{1}{R_{a.83}}\right)^{-1}\right]$ 　　$R_{o.a.83} = 38.462 \times 10^3 \, \Omega$

$R_{o.a.83.eff} := \left(\dfrac{1}{r_{a.83}} + \dfrac{1}{R_{a.83.eff}}\right)^{-1}$ 　　$R_{o.a.83.eff} = 37.037 \times 10^3 \, \Omega$

$r_{c.83} := \dfrac{r_{a.83} + R_{a.83}}{1 + \mu_{83}}$ 　　　　$r_{c.83} = 1.609 \times 10^3 \, \Omega$

$R_{o.c.83} := \left(\dfrac{1}{r_{c.83}} + \dfrac{1}{R_{c.83}}\right)^{-1}$ 　　　$R_{o.c.83} = 802.222 \times 10^0 \, \Omega$

$r_{c.83.eff} := \dfrac{r_{a.83} + R_{a.83.eff}}{1 + \mu_{83}}$ 　　　$r_{c.83.eff} = 1.519 \times 10^3 \, \Omega$

$R_{o.c.83.eff} := \left(\dfrac{1}{r_{c.83.eff}} + \dfrac{1}{R_{c.83}}\right)^{-1}$ 　　$R_{o.c.83.eff} = 779.198 \times 10^0 \, \Omega$

2.5.7　Capacitance C_c :

$f_c := 20 \text{Hz}$ 　　　　$f_{c.opt} := \dfrac{f_c}{100}$ 　　　$f_{c.opt} = 200 \times 10^{-3}$ Hz

$C_c := \dfrac{1}{2 \cdot \pi \cdot f_{c.opt} \cdot R_{o.c.83.eff}}$ 　　　　$C_c = 1.021 \times 10^{-3}$ F

2.5.8　Specific other capacitances:

$C_{i.tot.83} := \left(1 + \left|G_{83}\right|\right) \cdot C_{g.a.83} + C_{g.c.83} + C_{stray.83}$ 　　$C_{i.tot.83} = 111.712 \times 10^{-12}$ F

2.5.9   Gain stage frequency and phase response calculations:

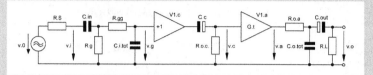

Figure 2.5 = Figure 2.3

$$Z1(f) := \frac{1}{2j \cdot \pi \cdot f \cdot C_{in.83}}$$

$$Z2(f) := \frac{1}{2j \cdot \pi \cdot f \cdot C_{i.tot.83}}$$

$$T_i(f) := \frac{Z2(f) \cdot \left(\dfrac{1}{R_{g.83}} + \dfrac{1}{R_{gg.83} + Z2(f)} \right)^{-1}}{\left(Z2(f) + R_{gg.83} \right) \cdot \left[R_S + Z1(f) + \left(\dfrac{1}{R_{g.83}} + \dfrac{1}{R_{gg.83} + Z2(f)} \right)^{-1} \right]}$$

$$\phi_i(f) := \operatorname{atan}\left(\frac{\operatorname{Im}\left(T_i(f) \right)}{\operatorname{Re}\left(T_i(f) \right)} \right)$$

$$T_{i.e}(f) := 20 \cdot \log\left(\left| T_i(f) \right| \right)$$

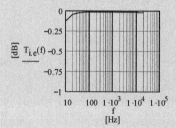

Figure 2.6 Transfer of i/p network

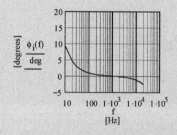

Figure 2.7 Phase of i/p network

$$Z3(f) := \frac{1}{2j \cdot \pi \cdot f \cdot C_{o.tot.83}}$$

$$Z4(f) := \frac{1}{2j \cdot \pi \cdot f \cdot C_{out.83}}$$

$$T_o(f) := \frac{\left(\dfrac{1}{Z3(f)} + \dfrac{1}{Z4(f) + R_L} \right)^{-1}}{R_{o.a.83} + \left(\dfrac{1}{Z3(f)} + \dfrac{1}{Z4(f) + R_L} \right)^{-1}} \cdot \frac{R_L}{R_L + Z4(f)}$$

$$\phi_o(f) := \operatorname{atan}\left(\frac{\operatorname{Im}\left(T_o(f) \right)}{\operatorname{Re}\left(T_o(f) \right)} \right)$$

$$T_{o.e}(f) := 20 \cdot \log\left(\left| T_o(f) \right| \right)$$

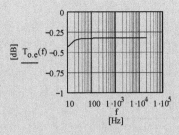

Figure 2.8 Transfer of cathode network

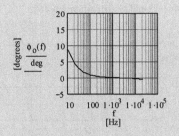

Figure 2.9 Phase of cathode network

$$T_c(f) := \left(\frac{R_{o.c.83.eff}}{R_{o.c.83.eff} + \dfrac{1}{2j \cdot \pi \cdot f \cdot C_c}} \right)$$

$$\phi_c(f) := atan\left(\frac{Im(T_c(f))}{Re(T_c(f))} \right)$$

$$T_{c.e}(f) := 20 \cdot log(|T_c(f)|)$$

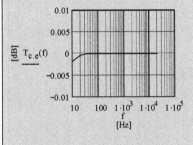

Figure 2.10 Transfer of o/p network

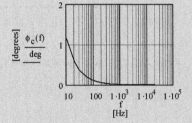

Figure 2.11 Phase of o/p network

$$T_{tot.83}(f) := T_i(f) \cdot T_o(f) \cdot T_c(f) \cdot G_{83}$$

$$\phi_{G.83}(f) := -180 \, deg$$

$$T_{tot.83.e}(f) := 20 \cdot log(|T_{tot.83}(f)|)$$

$$\phi_{tot.83}(f) := \phi_i(f) + \phi_o(f) + \phi_c(f) + \phi_{G.83}(f)$$

2.2.10 Frequency and phase response plots:

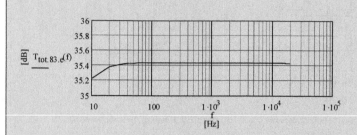

Figure 2.12 Frequency response of the whole CCS+Cc gain stage

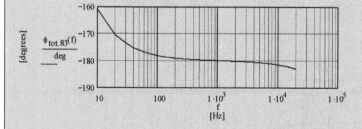

Figure 2.13 Phase response of the whole CCS+Cc gain stage

Chapter 3 The Cathode Follower (CF)

3.1   Circuit diagram - simple version CF1

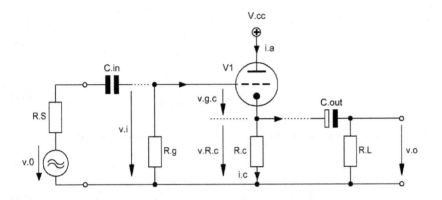

Figure 3.1 Basic design of a Cathode Follower gain stage (CF) with gain $G_{cfl} < 1$

3.2   Basic formulae (excl. stage load (R_L)-effect)

3.2.1 Gain G_{cf1} expressed in terms of $g_{m.t}$ and μ_t:

$$G_{cf1} = \frac{v_o}{v_i} \tag{3.1}$$

$$G_{cf1} = \frac{g_{m.t}\left(r_{a.t} \| R_c\right)}{1 + g_{m.t}\left(r_{a.t} \| R_c\right)} \tag{3.2}$$

$$\Rightarrow \quad G_{cf1} = \mu_t \frac{R_c}{r_{a.t} + \left(1 + \mu_t\right)R_c} \tag{3.3}$$

3.2.2 Grid input resistance R_g, input capacitance[1] $C_{i.tot}$ and input impedance $Z_{i.g}(f)$:

$$Z_{i.g}\left(f\right) = R_g \| C_{i.tot} \tag{3.4}$$

$$C_{i.tot} = C_{g.a} \| C_{stray} \| \left(1 - G_{cf1}\right)C_{g.c} \tag{3.5}$$

3.2.3 Cathode output resistance $R_{o.cl}$:

$$R_{o.cl} = r_{c.t} \| R_c = \frac{r_{c.t}\,R_c}{r_{c.t} + R_c} \tag{3.6}$$

$$r_{c.t} = \frac{r_{a.t}}{\mu_t + 1} \tag{3.7}$$

rot = rule of thumb: with $\mu_t \gg 1$ \hfill (3.8)

$r_{c.t.rot}$ becomes:

$$r_{c.t.rot} \approx \frac{1}{g_{m.t}} \tag{3.9}$$

Consequently, to get gains close to 1 $g_{m.t}$ or R_c must be increased.

$g_{m.t}$ always has certain limits set by the given valves on the market, thus, R_c will be the target to get gains close to 1 (see 3.2: CF - improved version, further down these pages).

---

[1]   see footnote 2 of Chapter 1

3.3   Derivations

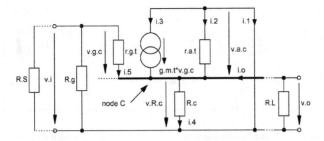

Figure 3.2   Equivalent circuit of Figure 3.1

$$v_{g.c} = v_i - v_{R.c}$$

$$v_{R.c} = v_o$$

$$v_o = -g_{m.t}\left(v_i - v_o\right)\left(R_c \parallel r_{a.t}\right)$$

$$\frac{v_o}{v_i} = g_{m.t}\left(1 - G_{cf1}\right)\left(R_c \parallel r_{a.t}\right) \tag{3.10}$$

$$G_{cf1} = g_{m.t}\frac{\left(R_c \parallel r_{a.t}\right)}{1 + g_{m.t}\left(R_c \parallel r_{a.t}\right)}$$

$$\mu_t = g_{m.t}\, r_{a.t}$$

$$G_{cf1} = \mu_t\frac{R_c}{r_{a.t} + \left(1 + \mu_t\right)R_c}$$

3.4   Gain stage frequency and phase response calculations

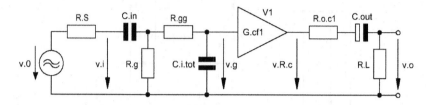

Figure 3.3 Simplified[2] equivalent circuit of Figure 3.1 - including all
frequency and phase response relevant components

3.4.1 Gain stage input transfer function T_{i1} (f) and phase φ_{i1} (f) - including source
resistance R_S and an oscillation preventing resistor $R_{gg} \ll R_g$:

from Chapter 1, paragraphs 1.4.1 & 1.4.2 we'll get T_i (f) and φ_i (f)[3] :

$$T_{i1}(f) = T_i(f)$$
$$\varphi_{i1}(f) = \varphi_i(f) \tag{3.11}$$

3.4.2 Gain stage output transfer function T_{o1} (f) and phase φ_{o1}(f):

$$T_{o1}(f) = \frac{v_o}{v_{R.c}} = \frac{R_L}{R_L + R_{o.c1} + Z4(f)} \tag{3.12}$$

$$Z4(f) = \left(2j\pi f\, C_{out}\right)^{-1} \tag{3.13}$$

$$\varphi_{o1}(f) = \arctan\left\{\frac{\mathrm{Im}\left[T_{o1}(f)\right]}{\mathrm{Re}\left[T_{o1}(f)\right]}\right\} \tag{3.14}$$

3.4.3 Fundamental gain stage phase shift $\varphi_{G.cf1.t}$(f):

$$\varphi_{G.cf1.t}(f) = 0° \tag{3.15}$$

3.4.4 Gain stage transfer function $T_{tot.1}$ (f) and phase $\varphi_{tot.1}$ (f):

$$T_{tot.1}(f) = T_{i1}(f)\, T_{o1}(f)\, G_{cf1} \tag{3.16}$$

$$\varphi_{tot.1}(f) = \varphi_{i1}(f) + \varphi_{o1}(f) + \varphi_{G.cf1}(f) \tag{3.17}$$

---

[2]   See footnote 2 of Chapter 1

[3]   to get $C_{i.tot}$ for the CF1 see equation 3.5 of this Chapter

3.5   Circuit diagram - improved version CF2$_b$ and CF2$_u$

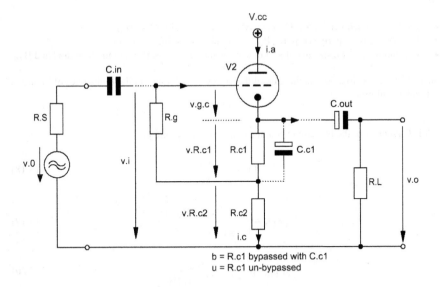

Figure 3.4 General CF gain stage with gain G$_{cf2}$ ≈ 1
and bootstrapped input resistance

3.6   Basic formulae (excl. stage load (R_L)-effect)

Assumed that

- for the bypassed version (b) C_{c1} is an AC short-circuit for R_{c1} in B_{20k}
- DC biasing is properly set by R_g, R_{c1}, R_{c2} to the A-class biasing point
- C_{out} and R_L are of a size that does not significantly influence the gain in the audio band B_{20k}

the basic formulae to calculate the gain of the Figure 3.4 CF look similar to the ones of Figure 3.1 and as follows:

3.6.1   Gain $G_{cf2.b}$ of the bypassed version (b) in terms of $g_{m.t}$ and μ_t:

$$G_{cf2.b} = \frac{v_o}{v_i} \tag{3.18}$$

$$G_{cf2.b} = \frac{g_{m.t}\left(r_{a.t} \parallel R_{c2}\right)}{1 + g_{m.t}\left(r_{a.t} \parallel R_{c2}\right)} \tag{3.19}$$

$$G_{cf2.b} = \mu_t \frac{R_{c2}}{r_{a.t} + \left(1 + \mu_t\right)R_{c2}} \tag{3.20}$$

3.6.2   Gain $G_{cf2.u}$ in case of absence of C_{c1} (= un-bypassed version u):

$$G_{cf2.u} = \frac{g_{m.t}\left[r_{a.t} \parallel \left(R_{c1} + R_{c2}\right)\right]}{1 + g_{m.t}\left[r_{a.t} \parallel \left(R_{c1} + R_{c2}\right)\right]} \tag{3.21}$$

$$G_{cf2.u} = \mu_t \frac{R_{c1} + R_{c2}}{r_{a.t} + \left(1 + \mu_t\right)\left(R_{c1} + R_{c2}\right)} \tag{3.22}$$

3.6.3   Input resistance $R_{i.g.2}$ and input impedance $Z_{i.g.2}(f)$:

$$R_{i.g.2.b} = \frac{R_g}{1 - G_{cf2.b}} \tag{3.23}$$

$$Z_{i.g.2.b}\left(f\right) = R_{i.g.2.b} \parallel C_{i.tot} \tag{3.24}$$

$$R_{i.g.2.u} = \frac{R_g}{1 - G_{cf2.u}\dfrac{R_{c2}}{R_{c1} + R_{c2}}} \tag{3.25}$$

$$Z_{i.g.2.u}\left(f\right) = R_{i.g.2.u} \parallel C_{i.tot} \tag{3.26}$$

3.6.4 Cathode output resistances[4] $R_{o.c2.b}$ & $R_{o.c2.u}$:

$$r_{c.t} = \frac{r_{a.t}}{\mu_t + 1} \tag{3.27}$$

rule of thumb:
$$r_{c.t.rot} \approx \frac{1}{g_{m.t}} \tag{3.28}$$

$$R_{o.c2.b} = r_{c.t} \parallel R_{c2} \tag{3.29}$$

$$R_{o.c2.u} = r_{c.t} \parallel (R_{c1} + R_{c2}) \tag{3.30}$$

3.6.5 C_{c1} calculation:

The audio band B_{20k} spreads from 20Hz to 20kHz. To ensure a flat frequency response in B_{20k} (±0.1dB) as well as a phase response deviation of less than 1° at $f_c = 20$Hz the C_{c1} calculation should be based on a corner frequency $f_{c.opt}$ that is $f_c/100$:

$$C_{c1} = \frac{1}{2\pi f_{c.opt} (r_{c.t} \parallel R_{c1})} \tag{3.31}$$

$$f_{c.opt} = \frac{f_c}{100} \tag{3.32}$$

3.6.6 Total input capacitance $C_{i.tot}$

calculation relevant valve capacitances:

$C_{g.c}$ = grid-cathode capacitance
$C_{g.a}$ = grid-plate capacitance = Miller capacitance
C_{stray} = sum of several different capacitances (to be guessed)[5]

$$C_{i.tot} = C_{g.a} \parallel C_{stray} \parallel (1 - G_{cf}) C_{g.c} \tag{3.33}$$

---

[4]   detailed derivation of $r_{c.t}$ see Chapter 4 "The Common Grid Stage (CGS)"
[5]   see footnote 2 of Chapter 1

3.7   Derivations

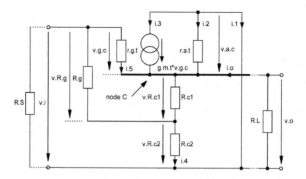

Figure 3.5 Equivalent circuit of Figure 3.4 (without C_{c1})

3.7.1 Gain $G_{cf2.b}$ (bypassed version $R_{c1} = 0$ for AC):

$$R_{c1} = 0$$

$$G_{cf2.b} = \frac{v_o}{v_i}$$

$$v_o = g_{m.t} v_{g.c} \left(r_{a.t} \parallel R_{c2} \right) \tag{3.34}$$

$$= v_{R.c2}$$

$$v_i = v_{R.g} + v_{R.c2}$$

$$= v_{g.c} + g_{m.t} v_{g.c} \left(r_{a.t} \parallel R_2 \right)$$

$$\Rightarrow G_{cf2.b} = \frac{g_{m.t} \left(r_{a.t} \parallel R_{c2} \right)}{1 + g_{m.t} \left(r_{a.t} \parallel R_{c2} \right)} \tag{3.35}$$

3.7.2 Gain $G_{cf2.u}$ (un-bypassed version):

$$G_{cf2.u} = \frac{v_o}{v_i}$$

$$v_o = g_{m.t} v_{g.c} \left(R_{c1} + R_{c2} \right) \tag{3.36}$$

$$v_i = v_{g.c} + g_{m.t} v_{g.c} (R_{c1} + R_{c2})$$

$$\Rightarrow G_{cf2.u} = \frac{g_{m.t} \left[r_{a.t} \parallel \left(R_{c1} + R_{c2} \right) \right]}{1 + g_{m.t} \left[r_{a.t} \parallel \left(R_{c1} + R_{c2} \right) \right]} \tag{3.37}$$

3.7.3 Input resistance $R_{i.g.b}$
(bypassed version [$R_{c1} = 0$ for AC] and its bootstrap effect):

$$i_5 = 0 \quad \Rightarrow \quad r_{g.t} = \infty$$

$$\Rightarrow \quad \frac{v_{g.c}}{R_g} = \frac{v_i}{R_{i.g.b}}$$

$$\frac{v_i - v_o}{R_g} = \frac{v_i}{R_{i.g.b}} \tag{3.38}$$

$$\frac{v_i \left(1 - G_{cf2.b}\right)}{R_g} = \frac{v_i}{R_{i.g.b}}$$

$$\Rightarrow \quad R_{i.g.b} = \frac{R_g}{1 - G_{cf2.b}} \tag{3.39}$$

3.7.4 Input resistance $R_{i.g.u}$
(un-bypassed version and its bootstrap effect):

$$\frac{v_{g.c} + v_{R.c1}}{R_g} = \frac{v_i}{R_{i.g.u}}$$

$$\frac{v_i \left(1 - G_{cf2.u}\right) + v_{R.c1}}{R_g} = \frac{v_i}{R_{i.g.u}} \tag{3.40}$$

$$v_{R.c1} = v_o \frac{R_{c1}}{R_{c1} + R_{c2}}$$

$$= v_i G_{cf2.u} \frac{R_{c1}}{R_{c1} + R_{c2}}$$

$$\Rightarrow \quad \frac{v_i \left(1 - G_{cf2.u}\right) + v_i \, G_{cf2.u} \dfrac{R_{c1}}{R_{c1} + R_{c2}}}{R_g} = \frac{v_i}{R_{i.g.u}} \tag{3.41}$$

$$\Rightarrow \quad R_{i.g.u} = \frac{R_g}{1 - G_{cf2.u} \dfrac{R_{c1}}{R_{c1} + R_{c2}}} \tag{3.42}$$

consequently:

$$R_{i.g.b} > \ldots \gg R_{i.g.u} \tag{3.43}$$

3.8   Gain stage frequency and phase response calculations

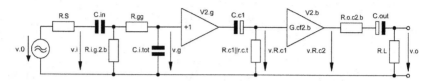

Figure 3.6 Simplified[6] equivalent circuit of Figure 3.3 (bypassed version) - including all frequency and phase response relevant components

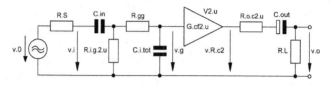

Figure 3.7 Simplified[7] equivalent circuit of Figure 3.3 (un-bypassed version) - including all frequency and phase response relevant components

3.8.1 Gain stage input transfer function T_{i1} (f) and phase φ_{i1} (f) - including source resistance R_S and an oscillation preventing resistor $R_{gg} \ll R_g$:

from Chapter 1, paragraphs 1.4.1 & 1.4.2 and with replacement of R_g by $R_{i.g.2.b}$ & $R_{i.g.2.u}$ we'll get T_i (f) and φ_i (f)[8] :

$$T_{i2.b}(f) = T_{i2.u}(f) = T_i(f)$$
$$\varphi_{i2.b}(f) = \varphi_{i2.u}(f) = \varphi_i(f)$$
(3.44)

3.8.2 Gain stage output transfer functions $T_{o2.b}$ (f), $T_{o2.u}$ (f) and phases $\varphi_{o2.b}(f)$, $\varphi_{o2.b}(f)$:

$$T_{o2.b}(f) = \frac{v_o}{v_{R.c2}} = \left(\frac{R_{c1} \| r_{c.t}}{R_{c1} \| r_{c.t} + Z3(f)} \right) \left(\frac{R_L}{R_L + R_{o.c2.b} + Z4(f)} \right)$$
(3.45)

$$Z3(f) = (2j\pi f \, C_{c1})^{-1}$$
(3.46)

$$Z4(f) = (2j\pi f \, C_{out})^{-1}$$
(3.47)

$$T_{o2.u}(f) = \frac{v_o}{v_{R.c2}} = \frac{R_L}{R_L + R_{o.c2.u} + Z4(f)}$$
(3.48)

---

6   see footnote 2 of Chapter 1 and footnote 3 of Chapter 2
7   see footnote 2 of Chapter 1
8   to get $C_{i.tot}$ for the CF2 see paragraph 3.6.6 of this Chapter

$$\varphi_{o2.b}(f)=\arctan\left\{\frac{\text{Im}\left[T_{o2.b}(f)\right]}{\text{Re}\left[T_{o2.b}(f)\right]}\right\} \qquad (3.49)$$

$$\varphi_{o2.u}(f)=\arctan\left\{\frac{\text{Im}\left[T_{o2.u}(f)\right]}{\text{Re}\left[T_{o2.u}(f)\right]}\right\} \qquad (3.50)$$

3.8.3 Fundamental gain stage phase shifts $\varphi_{G.cf2.b}(f)$ and $\varphi_{G.cf2.u}(f)$:

$$\varphi_{G.cf1.b}(f)=0° $$
$$\varphi_{G.cf1.u}(f)=0° \qquad (3.51)$$

3.8.4 Gain stage transfer functions $T_{tot.2.b}(f)$, $T_{tot.2.u}(f)$ and phases $\varphi_{tot.2.b}(f)$, $\varphi_{tot.2.u}(f)$:

$$T_{tot.2.u}(f)=T_{i2.u}(f)\,T_{o2.u}(f)\,G_{cf2.u} \qquad (3.52)$$

$$T_{tot.2.b}(f)=T_{i2.b}(f)\,T_{o2.b}(f)\,G_{cf2.b} \qquad (3.53)$$

$$\varphi_{tot.2.b}(f)=\varphi_{i2.b}(f)+\varphi_{o2.b}(f)+\varphi_{G.cf2.b}(f) \qquad (3.54)$$

$$\varphi_{tot.2.u}(f)=\varphi_{i2.u}(f)+\varphi_{o2.u}(f)+\varphi_{G.cf2.u}(f) \qquad (3.55)$$

3.9   CF1-Example with ECC83 / 12AX7 (83):

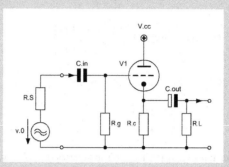

Figure 3.8 CF1 example circuitry

3.9.1   Triode bias data:

$I_{a.83} = 1.2 \cdot 10^{-3} A$ $\quad$ $V_{cc.83} = 250 V$ $\quad$ $V_{a.83} = 250 V$ $\quad$ $V_{g.83} = -2 V$

3.9.2   Triode valve constants:

$g_{m.83} := 1.6 \cdot 10^{-3} \cdot S$ $\qquad$ $\mu_{83} := 100$ $\qquad$ $r_{a.83} := 62.5 \cdot 10^{3} \Omega$

$C_{g.c.83} := 1.65 \cdot 10^{-12} F$ $\qquad$ $C_{g.a.83} := 1.6 \cdot 10^{-12} F$ $\qquad$ $C_{a.c.83} := 0.33 \cdot 10^{-12} F$

$\qquad\qquad\qquad\qquad\qquad\qquad\qquad\qquad\qquad\qquad\qquad\qquad$ $C_{o.83} := C_{a.c.83}$

3.9.3   Circuit variables:

$R_{a.83} := 0.001 \Omega$ $\qquad$ $R_{c.83} := 1.6 \cdot 10^{3} \Omega$ $\qquad$ $R_{g.83} := 1 \cdot 10^{6} \Omega$

$R_{S} := 1 \cdot 10^{3} \Omega$ $\qquad$ $R_{L} := 1 \cdot 10^{4} \Omega$ $\qquad$ $R_{gg.83} := 0.22 \cdot 10^{3} \Omega$

$C_{stray.83} := 10 \cdot 10^{-12} F$ $\qquad$ $C_{in.83} := 100 \cdot 10^{-9} F$ $\qquad$ $C_{out.83} := 100 \cdot 10^{-6} F$

3.9.4   Calculation relevant data:

frequency range f for the below shown graphs: $f := 10 Hz, 20 Hz .. 20000 Hz$

$\qquad\qquad\qquad\qquad\qquad\qquad\qquad\qquad\qquad\qquad\qquad\qquad$ $h := 1000 Hz$

3.9.5   Gain G_{cf1}:

$$G_{cf1.83} := \frac{g_{m.83} \cdot \left(\dfrac{r_{a.83} \cdot R_{c.83}}{r_{a.83} + R_{c.83}} \right)}{1 + g_{m.83} \cdot \left(\dfrac{r_{a.83} \cdot R_{c.83}}{r_{a.83} + R_{c.83}} \right)} \qquad\qquad G_{cf1.83} = 0.714$$

$$G_{cf1.83.e} := 20 \cdot \log\left(G_{cf1.83}\right) \qquad\qquad G_{cf1.83.e} = -2.926 \quad [dB]$$

3.9.6   Cathode output resistance $R_{o.c1}$:

$$r_{c.83} := \frac{r_{a.83}}{1 + \mu_{83}} \qquad\qquad r_{c.83} = 618.812 \ \Omega$$

$$R_{o.c1.83} := \left(\frac{1}{r_{c.83}} + \frac{1}{R_{c.83}}\right)^{-1} \qquad\qquad R_{o.c1.83} = 446.229 \ \Omega$$

$$r_{c.83.rot} := \frac{1}{g_{m.83}} \qquad\qquad r_{c.83.rot} = 625 \ \Omega$$

3.9.7   Total input capacitance $C_{i.tot}$:

$$C_{i.tot.83} := C_{g.a.83} + C_{stray.83} + \left(1 - G_{cf1.83}\right) \cdot C_{g.c.83} \qquad\qquad C_{i.tot.83} = 12.072 \times 10^{-12} F$$

3.9.8   Gain stage frequency and phase responses:

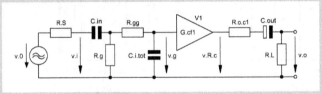

Figure 3.9 = Figure 3.3

$$Z1(f) := \frac{1}{2j \cdot \pi \cdot f \cdot C_{in.83}} \qquad\qquad Z2(f) := \frac{1}{2j \cdot \pi \cdot f \cdot C_{i.tot.83}}$$

$$T_{i1}(f) := \frac{Z2(f) \cdot \left(\dfrac{1}{R_{g.83}} + \dfrac{1}{R_{gg.83} + Z2(f)}\right)^{-1}}{\left(Z2(f) + R_{gg.83}\right) \cdot \left[R_S + Z1(f) + \left(\dfrac{1}{R_{g.83}} + \dfrac{1}{R_{gg.83} + Z2(f)}\right)^{-1}\right]}$$

$$T_{i1.e}(f) := 20 \cdot \log\left(\left|T_{i1}(f)\right|\right) \qquad\qquad \phi_{i1}(f) := atan\left(\frac{Im\left(T_{i1}(f)\right)}{Re\left(T_{i1}(f)\right)}\right)$$

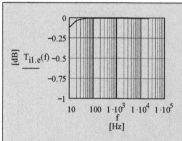

Figure 3.10 Transfer of i/p network

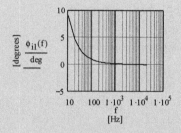

Figure 3.11 Phase of i/p network

$$Z4(f) := \frac{1}{2j \cdot \pi \cdot f \cdot C_{out.83}}$$

$$T_{o1}(f) := \frac{R_L}{R_L + R_{o.c1.83} + Z4(f)}$$

$$\phi_{o1}(f) := atan\left(\frac{Im(T_{o1}(f))}{Re(T_{o1}(f))}\right)$$

$$T_{o1.e}(f) := 20 \cdot log\left(\left|T_{o1}(f)\right|\right)$$

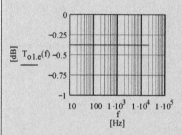

Figure 3.12 Transfer of o/p network

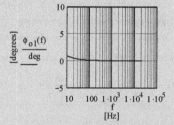

Figure 3.13 Phase of o/p network

$$T_{tot.1.83}(f) := T_{il}(f) \cdot T_{o1}(f) \cdot G_{cf1.83}$$

$$\phi_{G.cf1.83}(f) := 0 \cdot deg$$

$$T_{tot.1.83.e}(f) := 20 \cdot log\left(\left|T_{tot.1.83}(f)\right|\right)$$

$$\phi_{tot.1.83}(f) := \phi_{il}(f) + \phi_{o1}(f) + \phi_{G.cf1.83}(f)$$

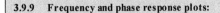

> ➤   MCD Worksheet III-1   CF1 calculations   Page 4

3.9.9   Frequency and phase response plots:

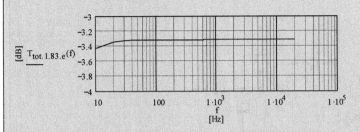

Figure 3.14 Frequency response of the whole CF1 gain stage

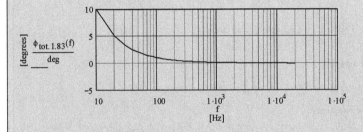

Figure 3.15 Phase response of the whole CF1 gain stage

3.10   CF2$_b$-Example with ECC83 / 12AX7 (83):

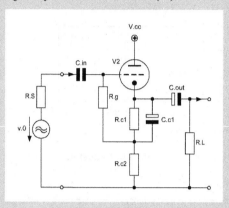

Figure 3.16 CF2$_b$ example circuitry

3.10.1 Triode bias data:

$$I_{a.83} = 1.2 \cdot 10^{-3}\,A \qquad V_{cc.83} = 370\,V \qquad V_{a.83} = 370\,V \qquad V_{g.83} = 122\,V$$

3.10.2 Triode valve constants:

$$g_{m.83} := 1.6 \cdot 10^{-3} \cdot S \qquad \mu_{83} := 100 \qquad r_{a.83} := 62.5 \cdot 10^{3}\,\Omega$$

$$C_{g.c.83} := 1.65 \cdot 10^{-12}\,F \qquad C_{g.a.83} := 1.6 \cdot 10^{-12}\,F \qquad C_{a.c.83} := 0.33 \cdot 10^{-12}\,F$$

$$C_{o.83} := C_{a.c.83}$$

3.10.3 Circuit variables:

$$R_{a.83} := 0.001\,\Omega \qquad R_{g.83} := 1 \cdot 10^{6}\,\Omega \qquad R_{c1.83} := 1.6 \cdot 10^{3}\,\Omega$$

$$R_S := 1 \cdot 10^{3}\,\Omega \qquad R_L := 10 \cdot 10^{3}\,\Omega \qquad R_{c2.83} := 100 \cdot 10^{3}\,\Omega$$

$$C_{stray.83} := 10 \cdot 10^{-12}\,F \qquad C_{in.83} := 100 \cdot 10^{-9}\,F \qquad C_{out.83} := 100 \cdot 10^{-6}\,F$$

$$R_{gg.83} := 0.22 \cdot 10^{3}\,\Omega$$

3.10.4 Calculation relevant data:

frequency range f for the below shown graphs: $f := 10\,Hz, 20\,Hz.. 20000\,Hz$

$h := 1000\,Hz$

> ➤   MCD Worksheet III-2   CF2$_b$ calculations   Page 2

3.10.5 Gain G$_{cf2.b}$:

$$G_{cf2.b.83} := \frac{g_{m.83} \cdot \left(\dfrac{r_{a.83} \cdot R_{c2.83}}{r_{a.83} + R_{c2.83}} \right)}{1 + g_{m.83} \cdot \left(\dfrac{r_{a.83} \cdot R_{c2.83}}{r_{a.83} + R_{c2.83}} \right)}$$

$$G_{cf2.b.83} = 0.984$$

$$G_{cf2.b.83.e} := 20 \cdot \log\left(G_{cf2.b.83}\right)$$

$$G_{cf2.b.83.e} = -0.14 \quad [dB]$$

3.10.6 Grid input resistance R$_{i.g2.b}$:

$$R_{i.g2.b.83} := \frac{R_{g.83}}{1 - G_{cf2.b.83}}$$

$$R_{i.g2.b.83} = 62.538 \times 10^6 \, \Omega$$

3.10.7 Cathode output resistance R$_{o.c2.b}$:

$$r_{c.83} := \frac{r_{a.83}}{1 + \mu_{83}}$$

$$r_{c.83} = 618.812 \, \Omega$$

$$R_{o.c2.b.83} := \left(\frac{1}{r_{c.83}} + \frac{1}{R_{c2.83}} \right)^{-1}$$

$$R_{o.c2.b.83} = 615.006 \, \Omega$$

$$r_{c.83.rot} := \frac{1}{g_{m.83}}$$

$$r_{c.83.rot} = 625 \, \Omega$$

3.10.8 C$_{c1}$ calculation:

$$f_c := 20\,Hz \qquad\qquad f_{c.opt} := \frac{f_c}{100} \qquad\qquad f_{c.opt} = 0.2\,Hz$$

$$C_{c1.83} := \frac{1}{2 \cdot \pi \cdot f_{c.opt} \cdot R_{o.c2.b.83}}$$

$$C_{c1.83} = 1.294 \times 10^{-3}\,F$$

3.10.9 Total input capacitance C$_{i.tot}$:

$$C_{i.tot.83} := C_{g.a.83} + C_{stray.83} + \left(1 - G_{cf2.b.83}\right) \cdot C_{g.c.83}$$

$$C_{i.tot.83} = 11.626 \times 10^{-12}\,F$$

3.10.10 Gain stage frequency and phase response :

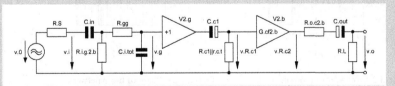

Figure 3.17 = Figure 3.6

$$Z1(f) := \frac{1}{2j \cdot \pi \cdot f \cdot C_{in.83}} \qquad\qquad Z2(f) := \frac{1}{2j \cdot \pi \cdot f \cdot C_{i.tot.83}}$$

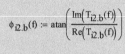

$$T_{i2.b}(f) := \frac{Z2(f) \cdot \left(\dfrac{1}{R_{i.g.2.b.83}} + \dfrac{1}{R_{gg.83} + Z2(f)} \right)^{-1}}{\left(Z2(f) + R_{gg.83}\right) \cdot \left[R_S + Z1(f) + \left(\dfrac{1}{R_{i.g.2.b.83}} + \dfrac{1}{R_{gg.83} + Z2(f)} \right)^{-1} \right]}$$

$$T_{i2.b.e}(f) := 20 \cdot \log\left(\left| T_{i2.b}(f) \right| \right) \qquad\qquad \phi_{i2.b}(f) := \operatorname{atan}\left(\frac{\operatorname{Im}\left(T_{i2.b}(f)\right)}{\operatorname{Re}\left(T_{i2.b}(f)\right)} \right)$$

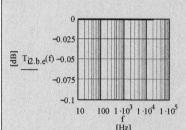

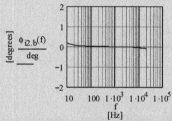

Figure 3.18 Transfer of b-i/p network   Figure 3.19 Phase of b-i/p network

$$Z3(f) := \frac{1}{2j \cdot \pi \cdot f \cdot C_{c1.83}} \qquad\qquad Z4(f) := \frac{1}{2j \cdot \pi \cdot f \cdot C_{out.83}}$$

$$T_{o2.b}(f) := \left(\frac{R_L}{R_L + R_{o.c2.b.83} + Z4(f)} \right) \cdot \left(\frac{R_{c1.83}}{R_{c1.83} + Z3(f)} \right)$$

$$T_{o2.b.e}(f) := 20 \cdot \log\left(\left| T_{o2.b}(f) \right| \right) \qquad\qquad \phi_{o2.b}(f) := \operatorname{atan}\left(\frac{\operatorname{Im}\left(T_{o2.b}(f)\right)}{\operatorname{Re}\left(T_{o2.b}(f)\right)} \right)$$

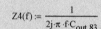

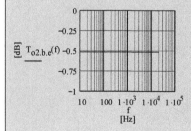

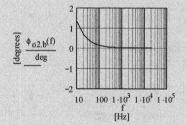

Figure 3.20 Transfer of b-o/p network   Figure 3.21 Phase of b-o/p network

$$T_{tot.2.b.83}(f) := T_{i2.b}(f) \cdot T_{o2.b}(f) \cdot G_{cf2.b.83} \qquad \phi_{G.cf2.b.83}(f) := 0 \cdot deg$$

$$T_{tot.2.b.83.e}(f) := 20 \cdot log\left(\left|T_{tot.2.b.83}(f)\right|\right) \qquad \phi_{tot.2.b.83}(f) := \phi_{i2.b}(f) + \phi_{o2.b}(f) + \phi_{G.cf2.b.83}(f)$$

3.10.11 Frequency and phase response plots:

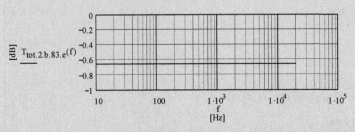

Figure 3.22 Frequency response of the whole CF2$_b$ gain stage

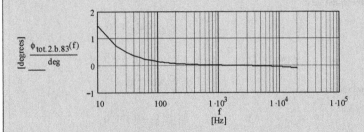

Figure 3.23 Phase response of the whole CF2$_b$ gain stage

3.11   CF2ᵤ-Example with ECC83 / 12AX7 (83):

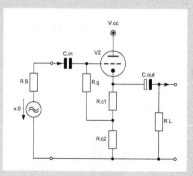

Figure 3.24 CF2ᵤ example circuitry

3.11.1 Triode bias data:

$$I_{a.83} = 1.2 \cdot 10^{-3} A \qquad V_{cc.83} = 370\,V \qquad V_{a.83} = 370\,V \qquad V_{g.83} = 122\,V$$

3.11.2 Triode valve constants:

$$g_{m.83} := 1.6 \cdot 10^{-3} \cdot S \qquad \mu_{83} := 100 \qquad r_{a.83} := 62.5 \cdot 10^{3}\,\Omega$$

$$C_{g.c.83} := 1.65 \cdot 10^{-12}F \qquad C_{g.a.83} := 1.6 \cdot 10^{-12}F \qquad C_{a.c.83} := 0.33 \cdot 10^{-12}F$$

$$C_{o.83} := C_{a.c.83}$$

3.11.3 Circuit variables:

$$R_{a.83} := 0.001\,\Omega \qquad R_{g.83} := 1 \cdot 10^{6}\,\Omega \qquad R_{c1.83} := 1.6 \cdot 10^{3}\,\Omega$$

$$R_{S} := 1 \cdot 10^{3}\,\Omega \qquad R_{L} := 1 \cdot 10^{4}\,\Omega \qquad R_{c2.83} := 100 \cdot 10^{3}\,\Omega$$

$$C_{stray.83} := 10 \cdot 10^{-12}F \qquad C_{in.83} := 100 \cdot 10^{-9}F \qquad C_{out.83} := 100 \cdot 10^{-6}F$$

$$R_{gg.83} := 0.22 \cdot 10^{3}\,\Omega$$

3.11.4 Calculation relevant data:

frequency range f for the below shown graphs: $f := 10\,Hz, 20\,Hz.. 20000\,Hz$

$$h := 1000\,Hz$$

3.11.5 Gain G_cf2.u :

$$G_{cf2.u.83} := \frac{g_{m.83}\left(\dfrac{1}{r_{a.83}} + \dfrac{1}{R_{c2.83} + R_{c1.83}}\right)^{-1}}{1 + g_{m.83}\left(\dfrac{1}{r_{a.83}} + \dfrac{1}{R_{c2.83} + R_{c1.83}}\right)^{-1}} \qquad G_{cf2.u.83} = 0.984$$

➤   MCD Worksheet III-3   CF2_u calculations   Page 2

$$G_{cf2.u.83.e} := 20 \cdot \log\left(G_{cf2.u.83}\right) \qquad\qquad G_{cf2.u.83.e} = -0.139 \quad [dB]$$

3.11.6 Grid input resistance $R_{i.g.2.u}$:

$$R_{i.g.2.u.83} := \frac{R_{g.83}}{1 - G_{cf2.u.83} \cdot \dfrac{R_{c2.83}}{R_{c2.83} + R_{c1.83}}} \qquad\qquad R_{i.g.2.u.83} = 31.855 \times 10^{6}\,\Omega$$

3.11.7 Cathode output resistance $R_{o.c2.u}$:

$$r_{c.83} := \frac{r_{a.83}}{1 + \mu_{83}} \qquad\qquad r_{c.83} = 618.812\,\Omega$$

$$R_{o.c2.u.83} := \left(\frac{1}{r_{c.83}} + \frac{1}{R_{c2.83} + R_{c1.83}}\right)^{-1} \qquad\qquad R_{o.c2.u.83} = 615.066\,\Omega$$

$$r_{c.83.rot} := \frac{1}{g_{m.83}} \qquad\qquad r_{c.83.rot} = 625\,\Omega$$

3.11.8 Total input capacitance $C_{i.tot}$:

$$C_{i.tot.83} := C_{g.a.83} + C_{stray.83} + \left(1 - G_{cf2.u.83}\right) \cdot C_{g.c.83} \qquad C_{i.tot.83} = 11.626 \times 10^{-12}\,F$$

3.11.9 Gain stage frequency and phase response:

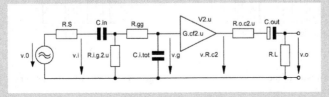

Figure 3.25 = Figure 3.7

$$Z1(f) := \frac{1}{2j \cdot \pi \cdot f \cdot C_{in.83}} \qquad\qquad\qquad Z2(f) := \frac{1}{2j \cdot \pi \cdot f \cdot C_{i.tot.83}}$$

$$T_{i2.u}(f) := \frac{Z2(f) \cdot \left(\dfrac{1}{R_{i.g.2.u.83}} + \dfrac{1}{R_{gg.83} + Z2(f)}\right)^{-1}}{\left(Z2(f) + R_{gg.83}\right) \cdot \left[R_S + Z1(f) + \left(\dfrac{1}{R_{i.g.2.u.83}} + \dfrac{1}{R_{gg.83} + Z2(f)}\right)^{-1}\right]}$$

$$T_{i2.u.e}(f) := 20 \cdot \log\left(\left|T_{i2.u}(f)\right|\right)$$

$$\phi_{i2.u}(f) := \operatorname{atan}\left(\frac{\operatorname{Im}\left(T_{i2.u}(f)\right)}{\operatorname{Re}\left(T_{i2.u}(f)\right)}\right)$$

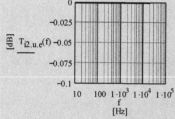

Figure 3.26 Transfer of u-i/p network   Figure 3.27 Phase of u-i/p network

$$T_{o2.u}(f) := \left(\frac{R_L}{R_L + R_{o.c2.u.83} + Z4(f)}\right)$$

$$Z4(f) := \frac{1}{2j \cdot \pi \cdot f \cdot C_{out.83}}$$

$$T_{o2.u.e}(f) := 20 \cdot \log\left(\left|T_{o2.u}(f)\right|\right)$$

$$\phi_{o2.u}(f) := \operatorname{atan}\left(\frac{\operatorname{Im}\left(T_{o2.u}(f)\right)}{\operatorname{Re}\left(T_{o2.u}(f)\right)}\right)$$

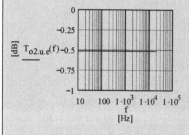

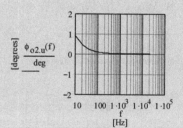

Figure 3.28 Transfer of u-o/p network   Figure 3.29 Phase of u-o/p network

$$T_{tot.2.u.83}(f) := T_{i2.u}(f) \cdot T_{o2.u}(f) \cdot G_{cf2.u.83} \qquad \phi_{G.cf2.u.83}(f) := 0 \cdot \deg$$

$$T_{tot.2.u.83.e}(f) := 20 \cdot \log\left(\left|T_{tot.2.u.83}(f)\right|\right) \qquad \phi_{tot.2.u.83}(f) := \phi_{i2.u}(f) + \phi_{o2.u}(f) + \phi_{G.cf2.u.83}(f)$$

3.11.10 Frequency and phase response plots:

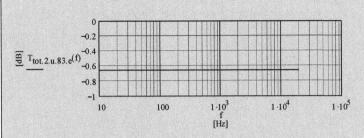

Figure 3.30 Frequency response of the whole CF2$_u$ gain stage

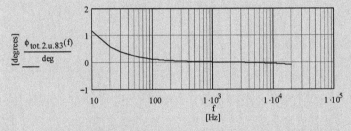

Figure 3.31 Phase response of the whole CF2$_u$ gain stage

4.1   Circuit diagram

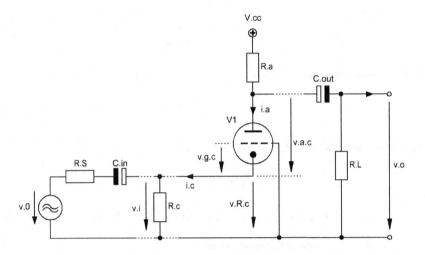

Figure 4.1 Basic design of a Common Grid Gain Stage (CGS)

4.2   Basic formulae (excl. stage load (R_L)-effect)

4.2.1 Gain G_t expressed in terms of $g_{m.t}$ and μ_t:

$$G_t = \left(g_{m.t}\, r_{a.t} + 1\right)\frac{R_a}{r_{a.t} + R_a} \tag{4.1}$$

$$G_t = \left(\mu_t + 1\right)\frac{R_a}{r_{a.t} + R_a} \tag{4.2}$$

4.2.2 Internal cathode input resistance $r_{c.t}$:

$$r_{c.t} = \frac{r_{a.t} + R_a}{\mu_t + 1} \tag{4.3}$$

4.2.3 Plate output resistance $R_{o.a}$ and impedance $Z_{o.a}(f)$:

$$R_{o.a} = R_a \parallel \left[r_{a.t} + \left(\mu_t + 1\right)\left(R_S \parallel R_c\right) \right] \tag{4.4}$$

$$Z_{o.a}(f) = R_a \parallel \left[r_{a.t} + \left(\mu_t + 1\right)\left(R_S \parallel R_c\right) \right] \parallel C_{o.tot} \tag{4.5}$$

4.2.4 Cathode input resistance $R_{i.c}$ and impedance $Z_{i.c}(f)$:

$$R_{i.c} = \frac{R_a + r_{a.t}}{\mu_t + 1} \parallel R_c \tag{4.6}$$

$$Z_{i.c}(f) = \frac{R_a + r_{a.t}}{\mu_t + 1} \parallel R_c \parallel C_{i.tot} \tag{4.7}$$

4.2.5 Input capacitance C_{in}:

$$C_{in} = \frac{1}{2\pi f_{c.opt}\, R_{i.c}} \tag{4.8}$$

$$f_{c.opt} = \frac{f_c}{100} = \frac{20\text{Hz}}{100} = 0.2\text{Hz} \tag{4.9}$$

4.3   Derivations

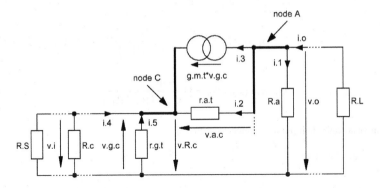

Figure 4.2 Equivalent circuit of Figure 4.1

4.3.1 Gain G_t and $G_{t.eff}$:

sum of currents at node A:

$$\frac{v_o}{R_a} + \frac{v_o - v_i}{r_{a.t}} + g_{m.t}\, v_{g.c} = i_o \qquad (4.10)$$

AC voltage situation at node C:

$$v_{g.c} = -v_i \qquad (4.11)$$

we set $i_0 = 0$ and do not take into account R_L, R_S and R_c, hence,

$$\frac{v_o}{R_a} + \frac{v_o - v_i}{r_{a.t}} = g_{m.t}\, v_i \qquad (4.12)$$

rearrangement leads to the stage gain G_t:

$$G_t = \frac{v_o}{v_i} = \frac{g_{m.t} + \dfrac{1}{r_{a.t}}}{\dfrac{1}{r_{a.t}} + \dfrac{1}{R_a}} \qquad (4.13)$$

$$\Rightarrow G_t = (\mu + 1)\frac{R_a}{r_{a.t} + R_a} \qquad (4.14)$$

$$\Rightarrow G_t = (g_{m.t}\, r_{a.t} + 1)\frac{R_a}{r_{a.t} + R_a} \qquad (4.15)$$

$R_{a.eff}$ replaces R_a in all calculations that include R_L as gain stage load!

$$R_{a.eff} = R_a \parallel R_L \qquad (4.16)$$

$$G_{t.eff} = (\mu + 1) \frac{R_a \parallel R_L}{r_{a.t} + (R_a \parallel R_L)} \qquad (4.17)$$

$$G_{t.eff} = (g_{m.t}\, r_{a.t} + 1) \frac{R_a \parallel R_L}{r_{a.t} + (R_a \parallel R_L)} \qquad (4.18)$$

4.3.2 Internal cathode input impedance $r_{c.t}$ and $r_{c.t.eff}$:

at node C we find:

$$r_{c.t} = \frac{v_i}{i_4} \qquad (4.19)$$

at node A we find:

$$i_1 = -i_4 = -(i_2 + i_3) \qquad (4.20)$$

thus,

$$i_1 = \frac{v_o}{R_a} = \frac{v_i}{r_{c.t}}$$
$$\Rightarrow \frac{v_o}{v_i} = G_t = \frac{R_a}{r_{c.t}} \qquad (4.21)$$

$$r_{c.t} = \frac{R_a}{G_t}$$
$$r_{c.t} = \frac{r_{a.t} + R_a}{\mu + 1} \qquad (4.22)$$

$$r_{c.t.eff} = \frac{R_{a.eff}}{G_{t.eff}} \qquad (4.23)$$

4.3.3 The inclusion of R_c and R_S into the calculation course:

$r_{c.t}$ becomes $r_{i.c.t.eff}$ the following way:

$$r_{i.c.t.eff} = r_{c.t} \parallel R_c \parallel R_S \qquad (4.24)$$

4.3.4 Total input and output capacitances $C_{i.tot}$ and $C_{o.tot}$:

As long as the valve is suitable for cascoded operation (eg. 6922 / E88CC, etc.) the input capacitance C_i of the cathode can be found in data sheets under cathode input capacitance C_i in CGS mode:

$$C_i = C_{c.g} \qquad (4.25)$$

hence, $C_{i.tot}$ becomes:

$$C_{i.tot} = C_i \,\|\, C_{stray} \qquad (4.26)$$

The data sheet output capacitance C_o in CGS mode becomes:

$$C_o = C_{a.g} \qquad (4.27)$$

hence, $C_{o.tot}$ becomes:

$$C_{o.tot} = C_{a.g} \,\|\, C_{a.c} \qquad (4.28)$$

Input and output capacitances in data sheets:

C_i = CGS input capacitance between cathode and grid   = $C_{c.g}$
       = CCS input capacitance between grid and cathode   = $C_{g.c}$

C_o = CGS output capacitance between plate and grid   = $C_{a.g}$
       = CCS output capacitance between plate and cathode   = $C_{a.c}$

4.4   Gain stage frequency and phase response

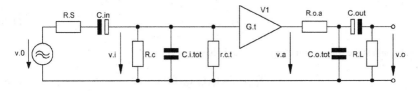

Figure 4.3 Simplified[1] equivalent circuit of Figure 4.1 - including all frequency and phase response relevant components

4.4.1 Gain stage input transfer function $T_i(f)$ and phase $\varphi_i(f)$ - including source resistance R_S:

$$T_i(f) = \frac{v_i}{v_0} \qquad (4.29)$$

---

[1]   see footnote 3 of Chapter 1

$$\varphi_i\left(f\right)=\arctan\left\{\frac{\text{Im}\left[T_i\left(f\right)\right]}{\text{Re}\left[T_i\left(f\right)\right]}\right\} \tag{4.30}$$

$$T_i\left(f\right)=\frac{\left[\left(\dfrac{1}{r_{c.t}}+\dfrac{1}{R_c}\right)+2j\pi f\,C_{i.tot}\right]^{-1}}{R_S+\dfrac{1}{2j\pi f\,C_{in}}+\left[\left(\dfrac{1}{r_{c.t}}+\dfrac{1}{R_c}\right)+2j\pi f\,C_{i.tot}\right]^{-1}} \tag{4.31}$$

4.4.2 Gain stage output transfer function $T_o(f)$:

$$T_o\left(f\right)=\frac{v_o}{v_a} \tag{4.32}$$

$$\varphi_o\left(f\right)=\arctan\left\{\frac{\text{Im}\left[T_o\left(f\right)\right]}{\text{Re}\left[T_o\left(f\right)\right]}\right\} \tag{4.33}$$

$$T_o\left(f\right)=\left(\frac{Z1(f)\,\|\,\left(Z2(f)+R_L\right)}{R_{o.a}+\left[Z1(f)\,\|\,\left(Z2(f)+R_L\right)\right]}\right)\left(\frac{R_L}{R_L+Z2(f)}\right) \tag{4.34}$$

$$Z1(f)=\left(2j\pi f\,C_{o.tot}\right)^{-1}$$
$$Z2(f)=\left(2j\pi f\,C_{out}\right)^{-1} \tag{4.35}$$

4.4.3 Fundamental gain stage phase shift $\varphi_{G.t}(f)$:

$$\varphi_{G.t}\left(f\right)=0° \tag{4.36}$$

4.4.4 Gain stage transfer function $T_{tot}(f)$ and phase $\varphi_{tot}(f)$:

$$T_{tot}\left(f\right)=T_i(f)\,T_o(f)\,G_t \tag{4.37}$$

$$\varphi_{tot}\left(f\right)=\varphi_i(f)+\varphi_o(f)+\varphi_{G.t}(f) \tag{4.38}$$

4.5   1st example with EC92 / 6AB4 (92):

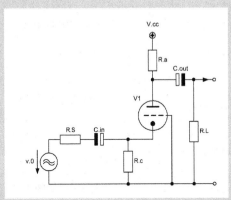

Figure 4.4 CGS example circuitry

4.5.1   Triode bias data:

$I_{a.92} = 10 \cdot 10^{-3} A$ $V_{cc.92} = 370 V$ $V_{a.92} = 250 V$ $V_{g.92} = -2V$

4.5.2   Triode valve constants:

$g_{m.92} := 5.6 \cdot 10^{-3} \cdot S$ $\mu_{92} := 60$ $r_{a.92} := 10.7 \cdot 10^3 \Omega$

$C_{i.92} := 4.6 \cdot 10^{-12} F$ $C_{g.a.92} := 1.8 \cdot 10^{-12} F$ $C_{o.92} := 2 \cdot 10^{-12} F$

4.5.3   Circuit variables:

$R_{a.92} := 12 \cdot 10^3 \Omega$ $R_{c.92} := 0.2 \cdot 10^3 \Omega$ $R_L := 1 \cdot 10^6 \Omega$

$R_S := 50 \Omega$ $C_{stray.92} := 10 \cdot 10^{-12} F$ $C_{out.92} := 1 \cdot 10^{-6} F$

4.5.4   Calculation relevant data:

frequency range f for the below shown graphs: $f := 10 Hz, 20 Hz .. 20000 Hz$

                                                        $h := 1000 Hz$

4.5.5   Gain G_t:

$G_{92} := (\mu_{92} + 1) \cdot \dfrac{R_{a.92}}{r_{a.92} + R_{a.92}}$ $G_{92} = 32.247$

$G_{92.e} := 20 \cdot \log(G_{92})$ $G_{92.e} = 30.17$ [dB]

4.5.6   Internal cathode input resistance $r_{c.t}$:

$$r_{c.92} := \frac{r_{a.92} + R_{a.92}}{\mu_{92} + 1}$$

$r_{c.92} = 372.131\,\Omega$

4.5.7   Plate output resistance $R_{o.a}$:

$$R_{o.a.92} := \left[\frac{1}{R_{a.92}} + \frac{1}{r_{a.92} + \left(\mu_{92} + 1\right)\cdot\left(\dfrac{1}{R_S} + \dfrac{1}{R_{c.92}}\right)^{-1}} \right]^{-1}$$

$R_{o.a.92} = 6.272 \times 10^3\,\Omega$

4.5.8   Cathode input resistance $R_{i.c}$:

$$R_{i.c.92} := \left(\frac{1}{R_{c.92}} + \frac{1}{r_{c.92}} \right)^{-1}$$

$R_{i.c.92} = 130.086\,\Omega$

4.5.9   Input capacitance C_{in}:

$$f_c := 20\,Hz \qquad\qquad f_{c.opt} := \frac{f_c}{100} \qquad\qquad f_{c.opt} = 0.2\,Hz$$

$$C_{in.92} := \frac{1}{2\cdot\pi\cdot f_{c.opt}\cdot R_{i.c.92}} \qquad\qquad C_{in.92} = 6.117 \times 10^{-3}\,F$$

4.5.10 Total input and output capacitance $C_{i.tot}$ and $C_{o.tot}$:

$$C_{i.tot.92} := C_{stray.92} + C_{i.92} \qquad\qquad C_{i.tot.92} = 14.6 \times 10^{-12}\,F$$

$$C_{o.tot.92} := C_{g.a.92} + C_{o.92} \qquad\qquad C_{o.tot.92} = 3.8 \times 10^{-12}\,F$$

4.5.11 Gain stage frequency and phase response:

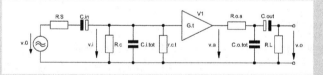

Figure 4.5 = Figure 4.3

$$T_i(f) := \frac{\left[\left(\dfrac{1}{r_{c.92}} + \dfrac{1}{R_{c.92}}\right) + 2j\cdot\pi\cdot f\cdot C_{i.tot.92}\right]^{-1}}{R_S + \dfrac{1}{2j\cdot\pi\cdot f\cdot C_{in.92}} + \left[\left(\dfrac{1}{r_{c.92}} + \dfrac{1}{R_{c.92}}\right) + 2j\cdot\pi\cdot f\cdot C_{i.tot.92}\right]^{-1}}$$

➢   MCD Worksheet IV   CGS calculations   Page 3

$$T_{i.e}(f) := 20 \cdot \log\left(\left|T_i(f)\right|\right)$$

$$\phi_i(f) := \operatorname{atan}\left(\frac{\operatorname{Im}\left(T_i(f)\right)}{\operatorname{Re}\left(T_i(f)\right)}\right)$$

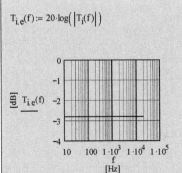

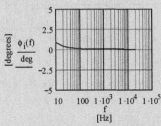

Figure 4.6 Transfer of i/p network   Figure 4.7 Phase of i/p network

$$Z1(f) := \left(2j \cdot \pi \cdot f \cdot C_{o.tot.92}\right)^{-1}$$

$$Z2(f) := \left(2j \cdot \pi \cdot f \cdot C_{out.92}\right)^{-1}$$

$$T_o(f) := \left[\frac{\left(\dfrac{1}{Z1(f)} + \dfrac{1}{Z2(f) + R_L}\right)^{-1}}{R_{o.a.92} + \left(\dfrac{1}{Z1(f)} + \dfrac{1}{Z2(f) + R_L}\right)^{-1}}\right] \cdot \left(\dfrac{R_L}{R_L + Z2(f)}\right)$$

$$T_{o.e}(f) := 20 \cdot \log\left(\left|T_o(f)\right|\right)$$

$$\phi_o(f) := \operatorname{atan}\left(\frac{\operatorname{Im}\left(T_o(f)\right)}{\operatorname{Re}\left(T_o(f)\right)}\right)$$

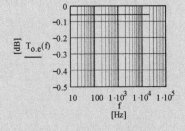

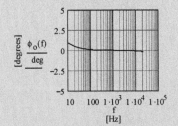

Figure 4.8 Transfer of o/p network   Figure 4.9 Phase of o/p network

$$T_{tot.92}(f) := T_i(f) \cdot T_o(f) \cdot G_{92}$$

$$\phi_{G.92}(f) := 0\,\mathrm{deg}$$

$$T_{tot.92.e}(f) := 20 \cdot \log\left(\left|T_{tot.92}(f)\right|\right)$$

$$\phi_{tot.92}(f) := \phi_i(f) + \phi_o(f) + \phi_{G.92}(f)$$

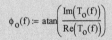

➤   MCD Worksheet IV   CGS calculations   Page 4

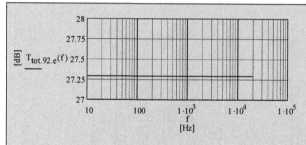

Figure 4.10 Frequency response of the whole CGS gain stage

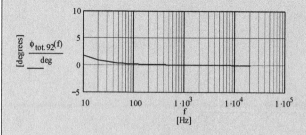

Figure 4.11 Phase response of the whole CGS gain stage

4.6   2nd example with EC92 / 6AB4 (92) with changed circuitry components and valve constants:

4.6.1   Triode bias data (n = new):

$I_{a.92.n} = 2 \cdot 10^{-3} A$ $V_{cc.92.n} = 370 V$ $V_{a.92.n} = 250 V$ $V_{g.92.n} = -4.4 V$

4.6.2   Triode valve constants:

$g_{m.92.n} := 1.6 \cdot 10^{-3} \cdot S$ $\mu_{92.n} := 45$ $r_{a.92.n} := 28 \cdot 10^{3} \Omega$

$C_{i.92.n} := 4.6 \cdot 10^{-12} F$ $C_{g.a.92.n} := 1.8 \cdot 10^{-12} F$ $C_{o.92.n} := 2 \cdot 10^{-12} F$

4.6.3   Circuit variables:

$R_{a.92.n} := 60 \cdot 10^{3} \Omega$ $R_{c.92.n} := 2.2 \cdot 10^{3} \Omega$ $R_L := 1 \cdot 10^{6} \Omega$

$R_S := 50 \Omega$

$C_{stray.92.n} := 10 \cdot 10^{-12} F$ $C_{out.92.n} := 1 \cdot 10^{-6} F$

4.6.4   Calculation relevant data:

frequency range f for the below shown graphs: $f := 10 Hz, 20 Hz .. 20000 Hz$

4.6.5   ff :

same calculation course as above leads to an overall gain improvement of appr. 2dB
- mainly because of a higher $R_{i.c.92}$ (130R1 vs. 1k023)

Chapter 5 Constant Current Generators (CCG)

Constant current generators as of Figure 5.1 ... 4 ensure stable DC current conditions for any kind of load impedance $R_{L1...4}$. This might be a resistance as well as another valve - including its associated components.

5.1   Constant Current Source (CCSo-lo) - low impedance version

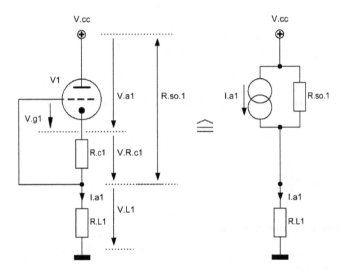

Figure 5.1 CCSo-lo that generates a lower impedance than the Figure 5.3 case

5.1.2 CCSo-lo:

With a rather low-valued R_{c1} (depending on the valve's bias settings) in Figure 5.1 a less complex version of the Figure 5.3 CCSo-hi leads to a lower valued Figure 5.1 resistance $R_{so.1}$ and impedance $Z_{so.1}(f)$.

Based on the derivation course and formulae found in Chapter 1.3 and the following equivalent circuit (see also Figure 1.2) $R_{so.1}$ and $Z_{so.1}(f)$ become:

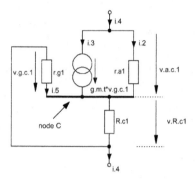

Figure 5.1-1 Equivalent circuit of the Figure 5.1 current source

$$R_{so.1} = r_{a.v1} + (1 + \mu_{v1}) R_{c1} \tag{5.1}$$

$$Z_{so.1}(f) = \left[(r_{a.v1} \parallel C_{a.c.1}) + (1 + \mu_{v1})(R_{c1} \parallel C_{g.c.1}) \right] \parallel C_{g.a.1} \parallel C_{stray.1} \tag{5.2}$$

5.1.2 DC voltage and current settings:

To get the required anode current I_{a1} R_{c1} sets the right gate voltage V_{g1} according to the following steps:

step 1   choose I_{a1}

step 2   choose V_{a1} at a given V_{cc}

step 3:   take V_{g1} from the valve's I_a/V_g vs. V_a characteristics diagram, hence,

$$V_{R.c1} = -V_{g1} \tag{5.3}$$

$$R_{c1} = \frac{|V_{g1}|}{I_{a1}} \tag{5.4}$$

thus, V_{L1} becomes:

$$V_{L1} = V_{cc} - V_{a1} - V_{R.c1} \tag{5.5}$$

5.2   Constant Current Sink (CCSi-lo) - low impedance version

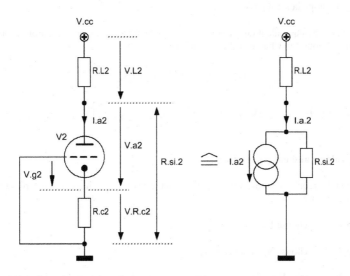

Figure 5.2 CCSi-lo that generates a lower impedance than the Figure 5.4 case

5.2.1 CCSi-lo:

With a rather low-valued R_{c2} in Figure 5.2 (depending on the valve's bias settings) a less complex version of the Figure 5.4 CCSi-hi leads to a lower valued Figure 5.2 resistance $R_{si.2}$ and impedance $Z_{si.2}(f)$.

Based on the equivalent circuit of the previous section and on the derivation course and formulae found in Chapter 1.3 $R_{si.2}$ and $Z_{si.2}(f)$ become:

$$R_{si.2} = r_{a.v2} + (1 + \mu_{v2}) R_{c2} \qquad (5.6)$$

$$Z_{si.2}(f) = \left[(r_{a.v2} \| C_{a.c.2}) + (1 + \mu_{v2})(R_{c2} \| C_{g.c.2}) \right] \| C_{g.a.2} \| C_{stray.2} \qquad (5.7)$$

5.2.2 DC voltage and current settings:

To get the required anode current I_{a2} R_{c2} sets the right gate voltage V_{g2} according to the following steps:

step 1   choose I_{a2}

step 2   choose V_{a2} at a given V_{cc}

step 3:   take V_{g2} from the valve's I_a/V_g vs. V_a characteristics diagram, hence,

$$V_{R.c2} = - V_{g2} \qquad (5.8)$$

$$R_{c2} = \frac{|V_{g2}|}{I_{a2}} \qquad (5.9)$$

thus, V_{L2} becomes:

$$V_{L2} = V_{cc} - V_{a2} - V_{R.c2} \qquad (5.10)$$

5.3 Constant Current Source (CCSo-hi) - high impedance version

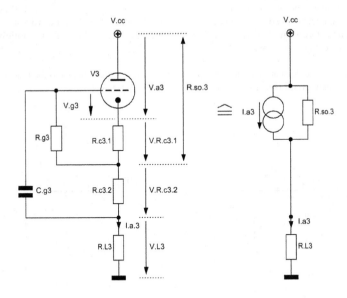

Figure 5.3 CCSo-hi that generates a higher impedance the Figure 5.1 case

5.3.1 CCSo-hi:

Basically, this rather complex CCG uses the same mechanics like the Figure 5.1 version. Additionally, because of the inclusion of C_{g3} the whole arrangement has to fulfil certain requirements concerning its frequency bandwidth. Otherwise, it would work like the Figure 5.1 version - but with much higher impedance $R_{so.3}$ and $Z_{so.3}(f)$. The following equivalent circuit shows why.

Based on the derivation course and formulae found in Chapter 1.3 and the below shown equivalent circuit (see also Figure 1.2) $R_{so.3}$ and $Z_{so.3}(f)$ become:

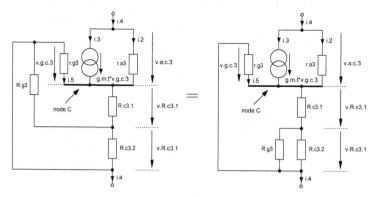

Figure 5.3-1 Equivalent circuit of Figure 5.3 - and its change
into a Figure 5.1 type of CCG

$$R_{so.3} = r_{a.v3} + (1 + \mu_{v3})(R_{c3.1} + R_{c3.2} \parallel R_{g3})$$ (5.11)

$$Z_{so.3}(f) = \left[(r_{a.v3} \parallel C_{a.c.3}) + (1 + \mu_{v3})(\{R_{c3.1} + R_{c3.2} \parallel R_{g3}\} \parallel C_{g.c.3}) \right] \parallel C_{g.a.3} \parallel C_{stray.3}$$ (5.12)

$\Rightarrow$ A high valued R_{c3} automatically leads to a very high valued $R_{so.3}$

$\Rightarrow$ A high valued $C_{stray.3}$ automatically leads to a heavy decrease of $Z_{so.3}(f)$ at
frequencies > 1kHz (see MCD Worksheet V-3)

5.3.2 DC voltage and current settings:

To get the required anode current I_{a3} R_{c3} sets the right gate voltage V_{g3} according to the following steps:

step 1   choose I_{a3}

step 2   choose V_{a3} at a given V_{cc}

step 3:   take V_{g3} from the valve's I_a/V_g vs. V_a characteristics diagram, hence,

$$V_{R.c3.1} = -V_{g3} \tag{5.13}$$

$$R_{c3.1} = \frac{|V_{g3}|}{I_{a3}} \tag{5.14}$$

step 4: choose $R_{c3.2}$, hence, $V_{R.c3.2}$ becomes:

$$V_{R.c3.2} = I_{a3} R_{c3.2} \tag{5.15}$$

step 5: calculate V_{L3}:

$$V_{L3} = V_{cc} - V_{a3} - V_{R.c3.1} - V_{R.c3.2} \tag{5.16}$$

! A voltage V_{L3} too low or too high has to lead to an adaptation of V_{cc} and / or $R_{c3.2}$!

step 6: calculate C_{g3}, the gain $G_{CCSo.hi}$ and $R_{in.3}$:

$$C_{g3} = \frac{1}{2\pi f_{c.opt} R_{in.3}} \tag{5.17}$$

$$f_c = 20\text{Hz}$$
$$f_{c.opt} = \frac{20\text{Hz}}{100} \tag{5.18}$$

Basically, the CCSo-hi equals a cathode follower type CF2$_u$. But its input is connected via the input capacitance C_{g3} to the bottom lead of $R_{c3.2}$, which acts as an input short-circuit for the CF2$_u$.

Consequently, $R_{in.3}$ can be calculated with the respective formulae of Chapter 3.6.2 ... 3.

$$R_{in.3} = \frac{R_{g3}}{1 - G_{CCSo.hi} \dfrac{\left(R_{c3.2} \parallel R_{g3}\right)}{R_{c3.1} + \left(R_{c3.2} \parallel R_{g3}\right)}} \tag{5.19}$$

$$G_{CCSo.hi} = \mu_{v3} \frac{R_{c3.1} + \left(R_{c3.2} \parallel R_{g3}\right)}{r_{a.v3} + \left(1 + \mu_{v3}\right)\left(R_{c3.1} + \left[R_{c3.2} \parallel R_{g3}\right]\right)} \tag{5.20}$$

It's obvious that at frequencies below $f_{c.opt}$ the impedance of C_{g3} will steadily increase, thus, turning this kind of CCG more and more into a CCSo-lo type of CCG.

5.4   Constant Current Sink (CCSi-hi) - high impedance version

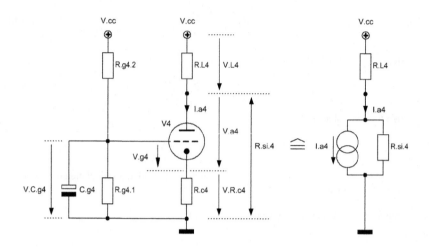

Figure 5.4 CCSi-hi that generates a higher impedance than the Figure 5.2 case

5.4.1 CCSi-hi:

Based on the equivalent circuit of the previous section and on the derivation course and formulae found in Chapter 1.3 $R_{si.4}$ and $Z_{si.4}(f)$ become:

$$R_{si.4} = r_{a.v4} + (1 + \mu_{v4}) R_{c4} \tag{5.21}$$

$$Z_{si.4}(f) = \left[(r_{a.v4} \parallel C_{a.c.4}) + (1 + \mu_{v4})(R_{c4} \parallel C_{g.c.4}) \right] \parallel C_{g.a.4} \parallel C_{stray.4} \tag{5.22}$$

$\Rightarrow$ A high valued R_{c4} (in conjunction with the right values for $R_{g4.1}$ and $R_{g4.2}$) automatically leads to a very high valued $R_{si.4}$.

$\Rightarrow$ A high valued $C_{stray.4}$ automatically leads to a heavy decrease of $Z_{so.4}(f)$ at frequencies > 1kHz (see MCD Worksheet V-4)

5.4.2 DC voltage and current settings:

To get the required anode current I_{a4} R_{c4}, $R_{g4.1}$ and $R_{g4.2}$ set the right gate voltage V_{g4} according to the following steps:

step 1:   choose I_{a4}

step 2:   choose V_{a4} at a given V_{cc}

step 3:   choose $V_{R.c4}$ and calculate R_{c4} and V_{L4}

$$R_{c4} = \frac{V_{R.c4}}{I_{a4}} \tag{5.23}$$

$$V_{L4} = V_{cc} - V_{a4} - V_{R.c4} \tag{5.24}$$

! A voltage V_{L4} too low or too high has to lead to an adaptation of V_{cc} and / or $V_{R.c4}$ and / or R_{c4} !

step 4:   take V_{g4} from the valve's I_a/V_g vs. V_a characteristics diagram and calculate $V_{C.g4}$

$$V_{C.g4} = V_{R.c4} + V_{g4} \tag{5.25}$$

step 5:   calculate voltage divider $R_{g4.1}$ and $R_{g4.2}$:

$$\frac{V_{C.g4}}{V_{cc}} = \frac{R_{g4.1}}{R_{g4.1} + R_{g4.2}} \tag{5.26}$$

$$\frac{R_{g4.2}}{R_{g4.1}} = \frac{V_{cc}}{V_{C.g4}} - 1 \tag{5.27}$$

$$R_{g4.1} \parallel R_{g4.2} \leq \text{max. allowed}$$
$$\text{grid resistance of the valve} \tag{5.28}$$

5.4.3 Calculation of C_{g4}:

$$C_{g4} = \frac{1}{2\pi f_{c.opt} \left(R_{g4.1} \parallel R_{g4.2} \right)} \tag{5.29}$$

$$f_c = 20\text{Hz}$$
$$f_{c.opt} = \frac{20\text{Hz}}{100} \tag{5.30}$$

$Z_{si.4}(f)$ wouldn't be touched negatively at the lower end of B_{20k} as long as C_{g4}'s value is high enough, hence, ensuring a flat frequency and phase response of <0.1dB and <1° at 20Hz. If this would not be the case, than, the noise voltage produced by $R_{g4.1} \parallel R_{g4.2}$ would become a growing negative factor at the output of the current sink. It would be multiplied with the valve's gain and added to the noise voltage of the valve. The amount of noise voltage depends on the lp (formed by C_{g4} and $R_{g4.1} \parallel R_{g4.2}$) corner frequency f_{noise} and on the size of the two resistors.

Therefore:

$$f_{noise} \leq f_{c.opt} \tag{5.31}$$

5.5   Constant current source (CCSo-lo)
Example with EC92 / 6AB4 (92):

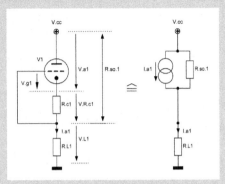

Figure 5.5 CCSo-lo example circuitry

5.5.1   DC voltage & current:

step 1 - choose I_{a1}:

$$I_{a1} := 5 \cdot 10^{-3} A$$

step 2 - choose V_{a1} at a given V_{cc}:

$$V_{a1} := 100 V \qquad\qquad V_{cc} := 370 V$$

step 3 - determine V_{g1}, calculate R_{c1} and V_{L1}:

$$V_{g1} := -0.7V \qquad\qquad V_{R.c1} := -V_{g1}$$

$$R_{c1} := \frac{|V_{g1}|}{I_{a1}} \qquad\qquad R_{c1} = 140\,\Omega$$

$$V_{L1} := V_{cc} - V_{a1} - V_{R.c1} \qquad\qquad V_{L1} = 269.3\ V$$

5.5.2   Impedances:

Determination of triode valve constants at I_{a1} and calculation of
$R_{so.1}$ and $Z_{so.1}(f)$:

$$g_{m.92} := 5 \cdot 10^{-3} \cdot S \qquad \mu_{92} := 66 \qquad r_{a.92} := 13.2 \cdot 10^{3}\,\Omega$$

$$C_{g.c.92} := 2.2 \cdot 10^{-12} F \qquad C_{g.a.92} := 1.5 \cdot 10^{-12} F \qquad C_{a.c.92} := 0.24 \cdot 10^{-12} F$$

$$C_{stray.1} := 10 \cdot 10^{-12} F \qquad X_{a.g.1}(f) := 2j \cdot \pi \cdot f \cdot \left(C_{g.a.92} + C_{stray.1} \right)$$

> **MCD Worksheet V-1**   CCSo-lo calculations   Page 2

Other calculation relevant data:

frequency range f for the below shown graphs:

$f := 10\text{Hz}, 20\text{Hz}.. \ 100000\text{Hz}$

$h := 1000 \cdot \text{Hz}$

$R_{so.1} := r_{a.92} + \left(1 + \mu_{92}\right) \cdot R_{c1}$ $R_{so.1} = 22.58 \times 10^3 \, \Omega$

$$Z_{so.1}(f) := \left[\left[\left(\frac{1}{r_{a.92}} + 2j \cdot \pi \cdot f \cdot C_{a.c.92}\right)^{-1} + \left(1 + \mu_{92}\right) \cdot \left[\frac{1}{R_{c1}} + \left(2j \cdot \pi \cdot f \cdot C_{g.c.92}\right)\right]^{-1}\right]^{-1} + X_{a.g.1}(f)\right]^{-1}$$

$\left|Z_{so.1}(h)\right| = 22.58 \times 10^3 \, \Omega$

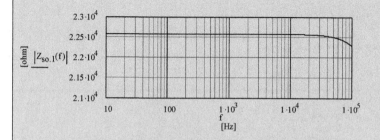

Figure 5.6 Frequency response of the impedance of the CCSo-lo

5.6 Constant current sink (CCSi-lo)
Example with EC92 / 6AB4 (92):

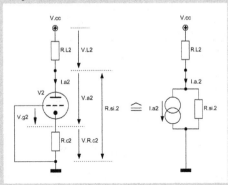

Figure 5.7 CCSi-lo example circuitry

5.6.1 DC voltage & current:

step 1 - choose I_{a2}:

$$I_{a2} := 5 \cdot 10^{-3} A$$

step 2 - choose V_{a2} at a given V_{cc}:

$$V_{a2} := 100 V \qquad\qquad V_{cc} := 370 V$$

step 3 - determine V_{g2}, calculate R_{c2} and V_{L2}:

$$V_{g2} := -0.7 V \qquad\qquad V_{R.c2} := -V_{g2}$$

$$R_{c2} := \frac{|V_{g2}|}{I_{a2}} \qquad\qquad R_{c2} = 140\,\Omega$$

$$V_{L2} := V_{cc} - V_{a2} - V_{R.c2} \qquad\qquad V_{L2} = 269.3 \text{ V}$$

5.6.2 Impedances:
Determination of triode valve constants at I_{a2} and calculation of
$R_{si.2}$ and $Z_{si.2}(f)$:

$$g_{m.92} := 5 \cdot 10^{-3} \cdot S \qquad \mu_{92} := 66 \qquad r_{a.92} := 13.2 \cdot 10^3 \Omega$$

$$C_{g.c.92} := 2.2 \cdot 10^{-12} F \qquad C_{g.a.92} := 1.5 \cdot 10^{-12} F \qquad C_{a.c.92} := 0.24 \cdot 10^{-12} F$$

$$C_{stray.2} := 10 \cdot 10^{-12} F \qquad X_{g.a.2}(f) := 2j \cdot \pi \cdot f \cdot \left(C_{g.a.92} + C_{stray.2} \right)$$

Other calculation relevant data:

frequency range f for the below shown graphs:

$f := 10\,Hz, 20\,Hz .. 100000\,Hz$

$h := 1000 \cdot Hz$

$R_{si.2} := r_{a.92} + (1 + \mu_{92}) \cdot R_{c2}$

$R_{si.2} = 22.58 \times 10^{3}\,\Omega$

$$Z_{si.2}(f) := \left[\left[\left(\frac{1}{r_{a.92}} + 2j \cdot \pi \cdot f C_{a.c.92} \right)^{-1} + (1 + \mu_{92}) \cdot \left[\frac{1}{R_{c2}} + \left(2j \cdot \pi \cdot f C_{g.c.92} \right) \right]^{-1} \right]^{-1} + X_{g.a.2}(f) \right]^{-1}$$

$\left| Z_{si.2}(h) \right| = 22.58 \times 10^{3}\,\Omega$

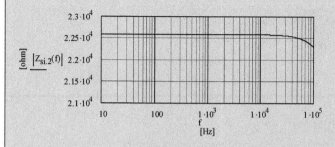

Figure 5.8 Frequency response of the impedance of the CCSi-lo

5.7   Constant current source (CCSo-hi)
Example with EC92 / 6AB4 (92):

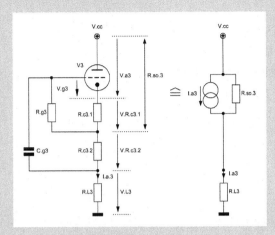

Figure 5.9 CCSo-hi example circuitry

5.7.1   DC voltage & current:

step 1 - choose I_{a3}:

$$I_{a3} := 5 \cdot 10^{-3} A$$

step 2 - choose V_{a3} at a given V_{cc}:

$$V_{a3} := 100 V \qquad\qquad\qquad V_{cc} := 370 V$$

step 3 - determine V_{g3}, calculate $R_{c3.1}$:

$$V_{g3} := -0.7 V$$

$$R_{c3.1} := \frac{|V_{g3}|}{I_{a3}} \qquad\qquad\qquad R_{c3.1} = 140 \times 10^{0} \Omega$$

step 4 - choose $R_{c3.2}$:

$$R_{c3.2} := 33.2 \cdot 10^{3} \Omega$$

$$V_{R.c3.2} := I_{a3} \cdot R_{c3.2} \qquad\qquad\qquad V_{R.c3.2} = 166 V$$

step 5 - calculate V_{L3}:

$$V_{R.c3.1} := -V_{g3}$$

$$V_{L3} := V_{cc} - V_{a3} - V_{R.c3.1} - V_{R.c3.2} \qquad\qquad V_{L3} = 103.3\,V$$

5.7.2   Impedances & capacitances:

5.7.2.1 Determination of triode valve constants at I_{a3}:

$$g_{m.92} := 5 \cdot 10^{-3} \cdot S \qquad\qquad \mu_{92} := 66 \qquad\qquad r_{a.92} := 13.2 \cdot 10^{3}\,\Omega$$

$$C_{g.c.92} := 2.2 \cdot 10^{-12}\,F \qquad C_{g.a.92} := 1.5 \cdot 10^{-12}\,F \qquad C_{a.c.92} := 0.24 \cdot 10^{-12}\,F$$

$$C_{stray.3} := 10 \cdot 10^{-12}\,F \qquad\qquad\qquad R_{g3} := 1 \cdot 10^{6}\,\Omega$$

Other calculation relevant data:

frequency range f for the below shown graphs:

$$f := 10Hz, 20Hz.. \ 100000\,Hz$$

$$h := 1000 \cdot Hz$$

5.7.2.2 step 6   Calculation of C_{c3}:

$$f_{c} := 20Hz \qquad\qquad f_{c.opt} := f_{c} \cdot 0.01 \qquad\qquad f_{c.opt} = 0.2\,Hz$$

$$G_{CCSo.hi.3} := \mu_{92} \frac{R_{c3.1} + \left(\dfrac{1}{R_{c3.2}} + \dfrac{1}{R_{g3}}\right)^{-1}}{r_{a.92} + \left(1 + \mu_{92}\right) \cdot \left[R_{c3.1} + \left(\dfrac{1}{R_{c3.2}} + \dfrac{1}{R_{g3}}\right)^{-1}\right]} \qquad G_{CCSo.hi.3} = 0.979$$

$$R_{in.3} := \frac{R_{g3}}{1 - G_{CCSo.hi.3} \cdot \dfrac{\left(\dfrac{1}{R_{c3.2}} + \dfrac{1}{R_{g3}}\right)^{-1}}{\left[R_{c3.1} + \left(\dfrac{1}{R_{c3.2}} + \dfrac{1}{R_{g3}}\right)^{-1}\right]}} \qquad R_{in.3} = 3.976 \times 10^{7}\,\Omega$$

$$C_{g3} := \frac{1}{2 \cdot \pi \cdot f_{c.opt} \cdot R_{in.3}} \qquad\qquad\qquad C_{g3} = 20.013 \times 10^{-9}\,F$$

5.7.2.3 Calculation of $R_{so.3}$ and $Z_{so.3}(f)$:

$$R_{c3} := R_{c3.1} + \left(\frac{1}{R_{c3.2}} + \frac{1}{R_{g3}} \right)^{-1}$$

$$X_3(f) := 2j \cdot \pi \cdot f \left(C_{g.a.92} + C_{stray.3} \right)$$

$$R_{so.3} := r_{a.92} + \left(1 + \mu_{92} \right) \cdot R_{c3}$$

$$R_{so.3} = 2.176 \times 10^6 \, \Omega$$

$$Z_{so.3}(f) := \left[\left[\left(\frac{1}{r_{a.92}} + 2j \cdot \pi \cdot f \cdot C_{a.c.92} \right)^{-1} + \left(1 + \mu_{92} \right) \cdot \left(\frac{1}{R_{c3}} + 2j \cdot \pi \cdot f \cdot C_{g.c.92} \right)^{-1} \right]^{-1} + X_3(f) \right]^{-1}$$

$$\left| Z_{so.3}(h) \right| = 2.149 \times 10^6 \, \Omega$$

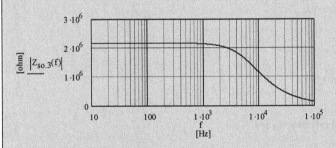

Figure 5.10 Frequency response of the impedance of the CCSo-hi

5.8   Constant current sink (CCSi-hi)
Example with EC92 / 6AB4 (92):

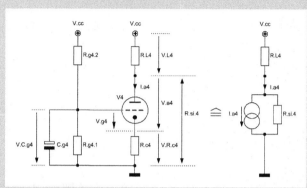

Figure 5.11 CCSi-hi example circuitry

5.8.1   DC voltage & current:

step 1 - choose I_{a4}:

$$I_{a4} := 5 \cdot 10^{-3} A$$

step 2 - choose V_{a4} at a given V_{cc}:

$$V_{a4} := 100 V \qquad\qquad\qquad V_{cc} := 370 V$$

step 3 - choose $V_{R.c4}$ and calculate R_{c4} and V_{L4}:

$$V_{R.c4} := 100 V$$

$$R_{c4} := \frac{V_{R.c4}}{I_{a4}} \qquad\qquad\qquad R_{c4} = 2 \times 10^4 \, \Omega$$

$$V_{L4} := V_{cc} - V_{a4} - V_{R.c4} \qquad\qquad V_{L4} = 170 \ V$$

step 4 - calculate $V_{C.g4}$:

$$V_{g4} := -0.7 V$$

$$V_{C.g4} := V_{R.c4} + V_{g4} \qquad\qquad V_{C.g4} = 99.3 \ V$$

step 5 - calculate $R_{g4.1}$ & $R_{g4.2}$:

$$R_{g4.1} := 1 \cdot 10^6 \Omega$$

$$R_{g4.2} := R_{g4.1} \cdot \left(\frac{V_{cc}}{V_{C.g4}} - 1 \right) \qquad\qquad R_{g4.2} = 2.726 \times 10^6 \, \Omega$$

> ➢   MCD Worksheet V-4   CCSi-hi calculations   Page 2

5.8.2 Impedances & capacitances:

5.8.2.1 Calculation of C_{g4}:

$$f_c := 20\,Hz \qquad\qquad f_{c.opt} := 0.01 \cdot 20\,Hz \qquad\qquad f_{c.opt} = 0.2\,Hz$$

$$C_{g4} := \frac{1}{2 \cdot \pi \cdot f_{c.opt} \left(\dfrac{1}{R_{g4.1}} + \dfrac{1}{R_{g4.2}} \right)^{-1}} \qquad\qquad C_{g4} = 1.088 \times 10^{-6}\,F$$

5.8.2.2 Determination of triode valve constants at I_{a4}:

$$g_{m.92} := 5 \cdot 10^{-3} \cdot S \qquad\qquad \mu_{92} := 66 \qquad\qquad r_{a.92} := 1.2 \cdot 10^{3}\,\Omega$$

$$C_{g.c.92} := 2.2 \cdot 10^{-12}\,F \qquad\qquad C_{g.a.92} := 1.4 \cdot 10^{-12}\,F \qquad\qquad C_{a.c.92} := 0.24 \cdot 10^{-12}\,F$$

$$C_{stray.4} := 10 \cdot 10^{-12}\,F \qquad\qquad X_{g.a.4}(f) := 2j \cdot \pi \cdot f \left(C_{g.a.92} + C_{stray.4} \right)$$

Other calculation relevant data:

frequency range f for the below shown graphs: $f := 10\,Hz, 20\,Hz .. 100000\,Hz$

$$h := 1000 \cdot Hz$$

5.8.2.3 Calculation of $R_{si.4}$ and $Z_{si.4}(f)$:

$$R_{si.4} := r_{a.92} + \left(1 + \mu_{92}\right) \cdot R_{c4} \qquad\qquad R_{si.4} = 1.341 \times 10^{6}\,\Omega$$

$$Z_{si.4}(f) := \left[\left[\left(\frac{1}{r_{a.92}} + 2j \cdot \pi \cdot f\,C_{a.c.92} \right)^{-1} + \left(1 + \mu_{92}\right) \left[\frac{1}{R_{c4}} + \left(2j \cdot \pi \cdot f\,C_{g.c.92}\right)^{-1} \right]^{-1} \right]^{-1} + X_{g.a.4}(f) \right]^{-1}$$

$$\left| Z_{si.4}(h) \right| = 1.335 \times 10^{6}\,\Omega$$

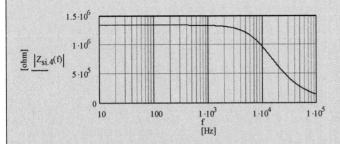

Figure 5.12 Frequency response of the impedance of the CCSi-hi

Chapter 6 The Cascode Amplifier (CAS)

6.1   Circuit diagram

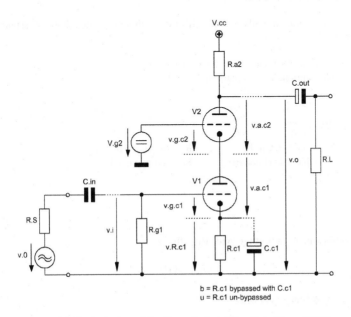

Figure 6.1 Basic design of the Cascode Amplifier Gain Stage (CAS)

6.2   Basic assumptions

Generally, the whole CAS gain stage consists of a CCS or CCS-Cc gain stage (V1 with gain G1) of Chapters 1 or 2 with a CGS gain stage (V2 with gain G2) as the plate load (alternatively, the gain of V1 can be increased by adding a resistor parallel to V2 plus $R_{a.2}$, thus, increasing the DC current of V1 as well).

$V_{g.2}$ sets the V2 biasing DC voltage, thus, defining the DC plate current I_a of both valves. $R_{c.1}$ has to be chosen adequately.

To get all relevant formulae shown further down these lines the following assumption were made:

- V1 equals V2 (ideal case: double triode)
- plate current V1   = plate current V2
- $g_{m.v1}$ $= g_{m.v2}$ $= g_{m.t}$
- $r_{a.v1}$ $= r_{a.v2}$ $= r_{a.t}$
- μ_{v1} $= \mu_{v2}$ $= \mu_t$
- R_{c2} $= R_{c1}$ $= R_c$
- bypassed version (b): R_{c1} = bypassed by C_{c1}
- un-bypassed version (u): R_{c1} = un-bypassed

6.3   Basic formulae (excl. stage load (R_L)-effect)

$$G_{cas} = G1\,G2 \tag{6.1}$$

6.3.1 bypassed version - gain $G_{cas.b}$:

$$G_{cas.b} = -\mu_t\,(\mu_t + 1)\frac{R_a}{(\mu_t + 2)\,r_{a.t} + R_a} \tag{6.2}$$

rule of thumb for $G_{cas.b}$:

$$G_{cas.b.rot} \approx -g_{m.t}\,R_a \tag{6.3}$$

6.3.2 un-bypassed version - gain $G_{cas.u}$:

$$G_{cas.u} = -\mu_t\,\frac{R_a + r_{a.t}}{R_a + r_{a.t}\,(\mu_t + 2) + R_{c1}\left(\mu_t^2 + \mu_t + 2\right)} \tag{6.4}$$

6.3.3 V2 plate output resistance $R_{o.a2}$ and impedance $Z_{o.a2}(f)$:

$$R_{o.a2.b} = \left(\frac{1}{R_a} + \frac{1}{r_{a.t}\,(\mu_t + 2)}\right)^{-1} \tag{6.5}$$

$$R_{o.a2.u} = \left(\frac{1}{R_a} + \frac{1}{r_{a.t}(\mu_t + 2) + (\mu_t + 1)^2 R_{cl}} \right)^{-1} \tag{6.6}$$

$$Z_{o.a2.b}(f) = R_a \parallel C_{a.g2} \parallel \left[(r_{a.t} \parallel C_{o.1}) + (\mu_t + 1)(r_{a.t} \parallel C_{o.2}) \right] \tag{6.7}$$

$$Z_{o.a2.u}(f) = R_a \parallel C_{a.g2} \parallel \left\{ \begin{array}{l} \left(\left[r_{a.v1} + (\mu_{v1} + 1) R_{cl} \right] \parallel C_{o.1} \right) + \\ \left[\left(r_{a.v2} + (\mu_{v2} + 1) \{ r_{a.v1} + (\mu_{v1} + 1) R_{cl} \} \right) \parallel C_{o.2} \right] \end{array} \right\} \tag{6.8}$$

6.4 Derivations

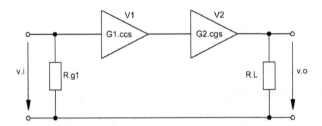

Figure 6.2 Simplified equivalent circuit of the CAS gain stage

6.4.1 Gains:

We do not need a rather complex equivalent circuit to describe the gain mechanics of the CAS gain stage. Without big extra derivation efforts all gains can be derived from Figure 6.2 plus the respective formulae given in the previous chapters.

$G1_b$ is a CCS+Cc gain stage with AC-grounded cathode à la Chapter 2. $G1_u$ is a CCS gain stage à la Chapter 1. G2 is a CGS gain stage à la Chapter 4. As the V1 plate load the internal cathode resistance of the CGS (V2) gain stage sets the gain G1.

6.4.1.1   bypassed version:

the internal cathode resistance $r_{c.v2}$ of the upper valve V2 becomes:

$$r_{c.v2} = \frac{r_{a.v2} + R_{a2}}{\mu_{v2} + 1} \tag{6.9}$$

the gain of a CCS+Cc gain stage becomes[1]:

---

[1]   see Chapter 2

$$G1_b = -g_{m.v1} \left(r_{a.v1} \parallel r_{c.v2} \right) \tag{6.10}$$

$$G1_b = -\mu_{v1} \frac{r_{c.v2}}{r_{a.v2} + r_{c.v2}} \tag{6.11}$$

thus, $G1_b$ becomes:

$$G1_b = -\mu_t \frac{r_{a.t} + R_a}{(\mu_t + 2) r_{a.t} + R_a} \tag{6.12}$$

G2 becomes[2]:

$$G2 = (\mu_t + 1) \frac{R_a}{r_{a.t} + R_a} \tag{6.13}$$

hence, $G_{cas.b}$ becomes:

$$
\begin{aligned}
G_{cas.b} &= G1_b \, G2 \\
&= -\mu_t (\mu_t + 1) \frac{R_a}{(\mu_t + 2) r_{a.t} + R_a}
\end{aligned} \tag{6.14}
$$

rule of thumb for $G_{cas.b}$:

$$
\begin{aligned}
G_{cas.b.rot} &\approx -\mu_t \frac{R_a}{r_{a.t} + \dfrac{R_a}{1 + \mu_t}} ; \frac{1}{1 + \mu_t} \ll 1 \\[2mm]
&\approx -\mu_t \frac{R_a}{r_{a.t}}
\end{aligned} \tag{6.15}
$$

$$G_{cas.b.rot} \approx -g_{m.t} R_a \tag{6.16}$$

6.4.1.2   un-bypassed version:

the gain of a CCS gain stage becomes[3]

$$G1_u = -\mu_{v1} \frac{r_{c.v2}}{r_{a.v1} + r_{c.v2} + (\mu_{v1} + 1) R_{c1}} \tag{6.17}$$

thus, $G1_u$ becomes:

$$G1_u = -\mu_t \frac{r_{a.t} + R_a}{r_{a.t} (\mu_t + 2) + R_a + (\mu_t + 2) R_{c1} + \mu_t^2 R_{c1}} \tag{6.18}$$

---

[2]   see Chapter 4
[3]   see Chapter 1

hence, $G_{cas.u}$ becomes:

$$G_{cas.u} = G1_u \, G2$$

$$= -\mu_t \frac{R_a + r_{a.t}}{R_a + r_{a.t}\left(\mu_t + 2\right) + R_{cl}\left(\mu_t^2 + \mu_t + 2\right)} \tag{6.19}$$

rule of thumb for $G_{cas.u}$: n. a.

6.4.2 Specific resistances and impedances:

6.4.2.1   bypassed version - output resistances and impedances at the plates of V1 and V2:

V1&V2 o/p resistance $R_{o.a1.b}$[4] and $R_{o.a2.b}$ and V2 o/p impedance $Z_{o.a2.b}(f)$:

$$R_{o.a1.b} = r_{a.v1} \tag{6.20}$$

According to Chapter 4 the o/p resistance $R_{o.a2.b}$ at the plate of V2 becomes:

$$R_{o.a2.b} = R_{a2} \parallel \left[r_{a.v2} + \left(\mu_{v2} + 1\right) r_{a.v1} \right] \tag{6.21}$$

$$R_{o.a2.b} = R_a \parallel \left[r_{a.t}\left(\mu_t + 2\right) \right] \tag{6.22}$$

$$Z_{o.a2.b}\left(f\right) = R_a \parallel C_{a.g2} \parallel \left[\left(r_{a.t} \parallel C_{o.1}\right) + \left(\mu_t + 1\right)\left(r_{a.t} \parallel C_{o.2}\right) \right] \tag{6.23}$$

since $C_{o.1} = C_{o.2} = C_o$ $Z_{o.a2.b}(f)$ becomes:

$$Z_{o.a2.b}\left(f\right) = R_a \parallel \left[\left(r_{a.t} \parallel C_o\right)\left(\mu_t + 2\right) \right] \tag{6.24}$$

6.4.2.2   un-bypassed version - output resistances and impedances at the plates of V1 and V2:

V1&V2 o/p resistance $R_{o.a1.u}$[5] and $R_{o.a2.u}$ and V2 o/p impedance $Z_{o.a2.u}(f)$:

$$R_{o.a1.u} = r_{a.v1} + \left(1 + \mu_{v1}\right) R_{cl} \tag{6.25}$$

According to Chapter 4 and with the respective equation from Chapter 1 $R_{o.a2.u}$ becomes:

$$R_{o.a2.u} = R_{a2} \parallel \left[r_{a.v2} + \left(\mu_{v2} + 1\right) r_{o.a.v1} \right] \tag{6.26}$$

---

[4]   see Chapter 2
[5]   see Chapter 1

$$r_{o.a.v1} = r_{a.v1} + (\mu_{v1} + 1) R_{cl} \tag{6.27}$$

$$R_{o.a2.u} = R_a \parallel \left[r_{a.t} (\mu_t + 2) + (\mu_t + 1)^2 R_{cl} \right] \tag{6.28}$$

$$Z_{o.a2.u} (f) = R_a \parallel C_{a.g2} \parallel \left\{ \begin{array}{l} \left(\left[r_{a.v1} + (\mu_{v1} + 1) R_{cl} \right] \parallel C_{o.1} \right) + \\ \left[\left(r_{a.v2} + (\mu_{v2} + 1) \left\{ r_{a.v1} + (\mu_{v1} + 1) R_{cl} \right\} \right) \parallel C_{o.2} \right] \end{array} \right\} \tag{6.29}$$

since $C_{o.1} = C_{o.2} = C_o$ $Z_{o.a2.u}(f)$ becomes:

$$Z_{o.a2.u} (f) = R_a \parallel C_{a.g2} \parallel \left\{ \begin{array}{l} \left[r_{a.t} + (\mu_t + 1) R_{cl} \right] \parallel C_o + \\ \left[r_{a.t} (\mu_t + 2) + (\mu_t + 1)^2 R_{cl} \right] \parallel C_o \end{array} \right\} \tag{6.30}$$

6.4.3 Role of R_L:

In the above given formulae it is assumed that $R_L \gg R_{a2}$, thus, not disturbing significantly the above shown calculation course. But a relatively small value of R_L should lead to new calculations of all formulae by replacing R_a with $R_{a.eff}$:

$$R_{a2.eff} = \frac{R_{a2} R_L}{R_{a2} + R_L} \tag{6.31}$$

6.5   Total input capacitance $C_{i.tot}$, total output capacitance $C_{o.tot}$ and cathode capacitance $C_{c.1}$

6.5.1 Input capacitances:

One of the advantages of the CGS gain stage is the rather low value of the input capacitance because of the very low gain $|G1|$ of $V1$[6]. Hence[7],

$$C_{i.1.tot.b} = \left[\left(1+|G1_b|\right)C_{g.a.v1}\right] \| C_{i.v1} \| C_{stray.1} \qquad (6.32)$$

$$C_{i.1.tot.u} = \left[\left(1+|G1_u|\right)C_{g.a.v1}\right] \| C_{i.v1} \| C_{stray.1} \qquad (6.33)$$

$$C_{i.2.tot} = C_{i.v2} \| C_{stray.2} \qquad (6.34)$$

6.5.2 Total output capacitance $C_{o.tot}$:

$$C_{o.1.tot} = C_{o.1} \| C_{g.a1} \qquad (6.35)$$

$$C_{o.2.tot} = C_{o.2} \| C_{g.a2} \qquad (6.36)$$

6.5.3 Input and output capacitances in data sheets:

C_i	= CGS input capacitance between cathode and grid	= $C_{g.c}$
	= CCS input capacitance between grid and cathode	= $C_{g.c}$
C_o	= CGS output capacitance between plate and grid	= $C_{g.a}$
	= CCS output capacitance between plate and cathode	= $C_{a.c}$

6.5.4 Capacitance C_{c1}[8]:

C_{c1} becomes[9]:

$$C_{c1} = \frac{1}{2\pi f_{c.opt} R_{o.c1}} \qquad (6.37)$$

$$R_{o.c1} = r_{c1} \| R_{c1} \qquad (6.38)$$

$$r_{c1} = \frac{r_{a.v1} + r_{c2}}{\mu_{v1} + 1} \qquad (6.39)$$

$$r_{c2} = \frac{r_{a.v2} + R_{a2}}{\mu_{v2} + 1} \qquad (6.40)$$

---

[6]	see MCD Worksheet VI - 6.7.8
[7]	for $C_{stray.1}$ and $C_{stray.2}$ (also $C_{i.1.tot}$ and $C_{i.2.tot}$) see footnote 2 of Chapter 1
[8]	see MCD Worksheet VI - 6.7.7 & 6.7.8
[9]	see respective formulae of Chapters 2 and 4

6.6   Gain stage frequency and phase response calculations

6.6.1 Bypassed version

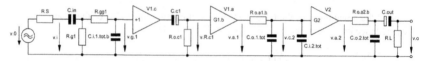

Figure 6.3 Simplified[10] equivalent circuit of Figure 6.1 (bypassed version)
- including all frequency and phase response relevant components

6.6.1.1   1st stage $(V1 = V1_c + V1_a)$ input transfer function $T_{i.1.b}(f)$ and phase $\varphi_{i.1.b}(f)$ -
including source resistance R_S and an oscillation preventing resistor $R_{gg1} \ll R_{g1}$:

$$T_{i.1.b}(f) = \frac{v_{g.1}}{v_0} \tag{6.41}$$

$$T_{i.1.b}(f) = \frac{Z2_b(f)\left(\dfrac{1}{R_{g1}} + \dfrac{1}{R_{gg1} + Z2_b(f)}\right)^{-1}}{\left(R_{gg1} + Z2_b(f)\right)\left[R_S + Z1(f) + \left(\dfrac{1}{R_{g1}} + \dfrac{1}{R_{gg1} + Z2_b(f)}\right)^{-1}\right]} \tag{6.42}$$

$$\begin{aligned} Z1(f) &= \left(2j\pi f\, C_{in}\right)^{-1} \\ Z2_b(f) &= \left(2j\pi f\, C_{i.1.tot.b}\right)^{-1} \end{aligned} \tag{6.43}$$

$$\varphi_{i.1.b}(f) = \arctan\left\{\frac{\operatorname{Im}\left[T_{i.1.b}(f)\right]}{\operatorname{Re}\left[T_{i.1.b}(f)\right]}\right\} \tag{6.44}$$

6.6.1.2   1st stage $V1_c$ cathode transfer function $T_{c.1.b}(f)$ and phase $\varphi_{c.1.b}(f)$:

$$T_{c.1.b}(f) = \frac{R_{o.c1}}{R_{o.c1} + \dfrac{1}{2j\pi f\, C_{c1}}} \tag{6.45}$$

$$\varphi_{c.1.b}(f) = \arctan\left\{\frac{\operatorname{Im}\left[T_{c.1.b}(f)\right]}{\operatorname{Re}\left[T_{c.1.b}(f)\right]}\right\} \tag{6.46}$$

---

10   concerning $C_{o.2}$ see footnote 3 of Chapter 1, concerning "simplified" see footnote 3 of
Chapter 2

6.6.1.3 1st stage output transfer function $T_{o.1.b}(f)$ and phase $\varphi_{o.1.b}(f)$:

$$T_{o.1.b}(f) = \frac{v_{c.2}}{v_{a.1}} \tag{6.47}$$

$$T_{o.1.b}(f) = \frac{Z3(f)}{R_{o.a1.b} + Z3(f)} \tag{6.48}$$

$$Z3(f) = \left[2j\pi f \left(C_{o.1.tot} + C_{i.2.tot} \right) \right]^{-1} \tag{6.49}$$

$$\varphi_{o.1.b}(f) = \arctan \left\{ \frac{Im\left[T_{o.1.b}(f) \right]}{Re\left[T_{o.1.b}(f) \right]} \right\} \tag{6.50}$$

6.6.1.4 2nd stage output transfer function $T_{o.2.b}(f)$ and phase $\varphi_{o.2.b}(f)$:

$$T_{o.2.b}(f) = \frac{v_o}{v_{a.2}} \tag{6.51}$$

$$T_{o.2.b}(f) = \left(\frac{Z4(f) \| \left(Z5(f) + R_L \right)}{R_{o.a.2.b} + \left[Z4(f) \| \left(Z5(f) + R_L \right) \right]} \right) \left(\frac{R_L}{R_L + Z5(f)} \right) \tag{6.52}$$

$$\begin{aligned} Z4(f) &= \left(2j\pi f \, C_{o.2.tot} \right)^{-1} \\ Z5(f) &= \left(2j\pi f \, C_{out} \right)^{-1} \end{aligned} \tag{6.53}$$

$$\varphi_{o.2.b}(f) = \arctan \left\{ \frac{Im\left[T_{o.2.b}(f) \right]}{Re\left[T_{o.2.b}(f) \right]} \right\} \tag{6.54}$$

6.6.1.5 1st stage fundamental phase shift $\varphi_{G.v1}(f)$:

$$\varphi_{G.v1}(f) = \varphi_{G.cas.b}(f) = -180° \tag{6.55}$$

6.6.1.6   total gain stage transfer function $T_{tot.b}(f)$ and phase $\varphi_{tot.b}(f)$:

$$T_{tot.b}(f) = T_{i.1.b}(f) T_{c.1.b}(f) T_{o.1.b}(f) T_{o.2.b}(f) G_{cas.b} \tag{6.56}$$

$$\varphi_{tot.b}(f) = \varphi_{i.1.b}(f) + \varphi_{c.1.b}(f) + \varphi_{o.1.b}(f) + \varphi_{o.2.b}(f) + \varphi_{G.cas.b}(f) \tag{6.57}$$

6.6.2 Un-bypassed version

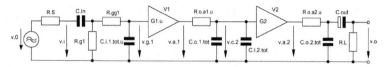

Figure 6.4 Simplified[11] equivalent circuit of Figure 6.1 (un-bypassed version)
- including all frequency and phase response relevant components

6.6.2.1   1st stage (V1) input transfer function $T_{i.1.u}(f)$ and phase $\varphi_{i.1.u}(f)$ -
             including source resistance R_S and an oscillation preventing resistor
             $R_{gg} \ll R_g$:

$$T_{i.1.u}(f) = \frac{v_{g.1}}{v_0} \tag{6.58}$$

$$T_{i.1.u}(f) = \frac{Z2_u(f)\left(\dfrac{1}{R_{g1}} + \dfrac{1}{R_{gg1} + Z2_u(f)}\right)^{-1}}{\left(R_{gg1} + Z2_u(f)\right)\left[R_S + Z1(f) + \left(\dfrac{1}{R_{g1}} + \dfrac{1}{R_{gg1} + Z2_u(f)}\right)^{-1}\right]} \tag{6.59}$$

$$Z1(f) = \left(2j\pi f\, C_{in}\right)^{-1}$$
$$Z2_u(f) = \left(2j\pi f\, C_{i.1.tot.u}\right)^{-1} \tag{6.60}$$

$$\varphi_{i.1.u}(f) = \arctan\left\{\frac{Im\left[T_{i.1.u}(f)\right]}{Re\left[T_{i.1.u}(f)\right]}\right\} \tag{6.61}$$

6.6.2.2   1st stage output transfer function $T_{o.1.u}(f)$ and phase $\varphi_{o.1.u}(f)$:

$$T_{o.1.u}(f) = \frac{v_{c.2}}{v_{a.1}} \tag{6.62}$$

$$T_{o.1.u}(f) = \frac{Z3(f)}{R_{o.a1.b} + Z3(f)} \tag{6.63}$$

$$Z3(f) = \left[2j\pi f\left(C_{o.1.tot} \parallel C_{i.2.tot}\right)\right]^{-1} \tag{6.64}$$

---

11   concerning $C_{o.1}$ and $C_{o.2}$ see footnote 3 of Chapter 1
        concerning $C_{i.1.tot}$ and $C_{i.2.tot}$ see footnote 2 of Chapter 1

$$\varphi_{o.1.u}(f) = \arctan\left\{\frac{\text{Im}\left[T_{o.1.u}(f)\right]}{\text{Re}\left[T_{o.1.u}(f)\right]}\right\} \tag{6.65}$$

6.6.2.3 2nd stage output transfer function $T_{o.2.u}(f)$ and phase $\varphi_{o.2.u}(f)$:

$$T_{o.2.u}(f) = \frac{v_o}{v_{a.2}} \tag{6.66}$$

$$T_{o.2.u}(f) = \left(\frac{Z4(f)\,\|\,\left(Z5(f)+R_L\right)}{R_{o.a2.u}+\left[Z4(f)\,\|\,\left(Z5(f)+R_L\right)\right]}\right)\left(\frac{R_L}{R_L+Z5(f)}\right) \tag{6.67}$$

$$Z4(f) = \left(2j\pi f\, C_{o.2.tot}\right)^{-1}$$
$$Z5(f) = \left(2j\pi f\, C_{out}\right)^{-1} \tag{6.68}$$

$$\varphi_{o.2.u}(f) = \arctan\left\{\frac{\text{Im}\left[T_{o.2.u}(f)\right]}{\text{Re}\left[T_{o.2.u}(f)\right]}\right\} \tag{6.69}$$

6.6.2.4 1st stage fundamental phase shift $\varphi_{G.v1}(f)$:

$$\varphi_{G.v1}(f) = \varphi_{G.cas.u}(f) = -180° \tag{6.70}$$

6.6.2.5   total gain stage transfer function $T_{tot.u}(f)$ and phase $\varphi_{tot.u}(f)$:

$$T_{tot.u}(f) = T_{i.1.u}(f)\,T_{o.1.u}(f)\,T_{o.2.u}(f)\,G_{cas.u} \tag{6.71}$$

$$\varphi_{tot.u}(f) = \varphi_{i.1.u}(f) + \varphi_{o.1.u}(f) + \varphi_{o.2.u}(f) + \varphi_{G.cas.u}(f) \tag{6.72}$$

6.7   Example with E188CC / 7308 (188):

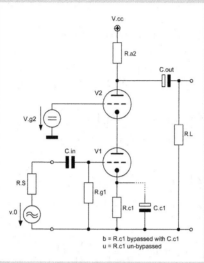

Figure 6.5 CAS example circuitry

6.7.1   Triode bias data:

$$I_{a1.188} = I_{a2.188} = 10 \cdot 10^{-3} A \qquad V_{a1.188} = 90V \qquad V_{a2.188} = 90V$$

$$V_{cc.188} = 370V \qquad V_{g1.188} = -1.65V \qquad V_{g2.188} = 90V$$

6.7.2   Triode valve constants:

$$g_{m.188} := 10.4 \cdot 10^{-3} \cdot S \qquad \mu_{188} := 32 \qquad r_{a.188} := 3.1 \cdot 10^3 \Omega$$

$$C_{g.c1.188} := 3.1 \cdot 10^{-12} F \qquad C_{g.a1.188} := 1.4 \cdot 10^{-12} F \qquad C_{a.c1.188} := 1.75 \cdot 10^{-12} F$$

$$C_{i.2.188} := 6 \cdot 10^{-12} F \qquad C_{g.a2.188} := 1.4 \cdot 10^{-12} F \qquad C_{a.c2.188} := 0.18 \cdot 10^{-12} F$$

$$C_{o.1.188} := C_{a.c1.188} \qquad\qquad\qquad C_{o.2.188} := 3 \cdot 10^{-12} F$$

6.7.3   Circuit variables:

$$R_{a2.188} := 12.667 \cdot 10^3 \Omega \qquad R_{c1.188} := 165\Omega \qquad R_{g1.188} := 47.5 \cdot 10^3 \Omega$$

$$R_S := 1 \cdot 10^3 \Omega \qquad R_L := 1 \cdot 10^6 \Omega \qquad R_{gg1.188} := 0.22 \cdot 10^3 \Omega$$

$$C_{stray.188.1} := 10 \cdot 10^{-12} F \qquad C_{in.188} := 10 \cdot 10^{-6} F \qquad C_{out.188} := 1 \cdot 10^{-6} F$$

$$C_{stray.188.2} := 10 \cdot 10^{-12} F$$

➤   MCD Worksheet VI   CAS calculations   Page 2

6.7.4   Calculation relevant data:

frequency range f for the below shown graphs: $f := 10\,Hz, 20\,Hz .. 100000\,Hz$

$h := 1000 \cdot Hz$

6.7.5   Gain $G_{cas.b}$:

$$G1_{188.b} := -\left[\mu_{188} \cdot \frac{r_{a.188} + R_{a2.188}}{(\mu_{188} + 2) \cdot r_{a.188} + R_{a2.188}} \right]$$
$G1_{188.b} = -4.273 \times 10^0$

$$G1_{188.b.e} := 20 \cdot \log\left(|G1_{188.b}| \right)$$
$G1_{188.b.e} = 12.615 \times 10^0$ [dB]

$$G2_{188} := (\mu_{188} + 1) \frac{R_{a2.188}}{r_{a.188} + R_{a2.188}}$$
$G2_{188} = 26.512 \times 10^0$

$$G2_{188.e} := 20 \cdot \log\left(|G2_{188}| \right)$$
$G2_{188.e} = 28.469 \times 10^0$ [dB]

$$G_{cas.188.b} := G1_{188.b} \cdot G2_{188}$$
$G_{cas.188.b} = -113.295 \times 10^0$

$$G_{cas.188.b.e} := 20 \cdot \log\left(|G_{cas.188.b}| \right)$$
$G_{cas.188.b.e} = 41.084 \times 10^0$ [dB]

$$G_{cas.188.b.rot} := -g_{m.188} \cdot R_{a2.188}$$
$G_{cas.188.b.rot} = -131.737 \times 10^0$

$$G_{cas.188.b.rot.e} := 20 \cdot \log\left(|G_{cas.188.b.rot}| \right)$$
$G_{cas.188.b.rot.e} = 42.394 \times 10^0$

$$G_{cas.188.b.e} - G_{cas.188.b.rot.e} = -1.31 \times 10^0 \quad \text{[dB]}$$

6.7.6   Gain $G_{cas.188.u}$:

$$G1_{188.u} := -\mu_{188} \cdot \frac{r_{a.188} + R_{a2.188}}{r_{a.188} \cdot (\mu_{188} + 2) + R_{a2.188} + (\mu_{188}^2 + \mu_{188} + 2) R_{c1.188}}$$

$G1_{188.u} = -1.724 \times 10^0$

$$G1_{188.u.e} := 20 \cdot \log\left(|G1_{188.u}| \right)$$
$G1_{188.u.e} = 4.731 \times 10^0$ [dB]

$$G_{cas.188.u} := G1_{188.u} \cdot G2_{188}$$
$G_{cas.188.u} = -45.71 \times 10^0$

$$G_{cas.188.u.e} := 20 \cdot \log\left(|G_{cas.188.u}| \right)$$
$G_{cas.188.u.e} = 33.2 \times 10^0$ [dB]

6.7.7   Specific resistances:

$$R_{o.a1.188.b} := r_{a.188}$$

$$r_{c2} := \frac{r_{a.188} + R_{a2.188}}{\mu_{188} + 1}$$
$r_{c2} = 477.788 \times 10^0 \, \Omega$

$$R_{o.c1} := \left[\left(\frac{r_{a.188} + r_{c2}}{\mu_{188} + 1}\right)^{-1} + \frac{1}{R_{c1.188}}\right]^{-1}$$ $$R_{o.c1} = 65.427 \times 10^0 \, \Omega$$

$$R_{o.a1.188.u} := r_{a.188} + (\mu_{188} + 1) \cdot R_{c1.188}$$ $$R_{o.a1.188.u} = 8.545 \times 10^3 \, \Omega$$

$$R_{o.a2.188.b} := \left[\frac{1}{r_{a.188} \cdot (\mu_{188} + 2)} + \frac{1}{R_{a2.188}}\right]^{-1}$$ $$R_{o.a2.188.b} = 11.308 \times 10^3 \, \Omega$$

$$R_{o.a2.188.u} := \left[\frac{1}{r_{a.188} \cdot (\mu_{188} + 2) + (\mu_{188} + 1)^2 \cdot R_{c1.188}} + \frac{1}{R_{a2.188}}\right]^{-1}$$

$$R_{o.a2.188.u} = 12.128 \times 10^3 \, \Omega$$

6.7.8   Specific input and output capacitances:

$$C_{i.1.tot.188.b} := \left(1 + |G1_{188.b}|\right) \cdot C_{g.a1.188} + C_{g.c1.188} + C_{stray.188.1}$$

$$C_{i.1.tot.188.u} := \left(1 + |G1_{188.u}|\right) \cdot C_{g.a1.188} + C_{g.c1.188} + C_{stray.188.1}$$

$$C_{i.1.tot.188.b} = 20.483 \times 10^{-12} F$$ $$C_{i.1.tot.188.u} = 16.914 \times 10^{-12} F$$

$$C_{i.2.tot.188} := C_{i.2.188} + C_{stray.188.2}$$ $$C_{i.2.tot.188} = 16 \times 10^{-12} F$$

$$C_{o.1.tot.188} := C_{o.1.188} + C_{g.a1.188}$$ $$C_{o.1.tot.188} = 3.15 \times 10^{-12} F$$

$$C_{o.2.tot.188} := C_{o.2.188} + C_{g.a2.188}$$ $$C_{o.2.tot.188} = 4.4 \times 10^{-12} F$$

$$f_c := 20 Hz$$ $$f_{c.opt} := \frac{f_c}{100}$$ $$f_{c.opt} = 200 \times 10^{-3} Hz$$

$$C_{c1} := \frac{1}{2 \cdot \pi \cdot f_{c.opt} \cdot R_{o.c1}}$$ $$C_{c1} = 12.163 \times 10^{-3} F$$

6.7.9   Gain stage frequency and phase response
- bypassed version:

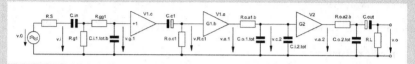

Figure 6.6 = Figure 6.3

$$Z1(f) := \frac{1}{2j \cdot \pi \cdot f \cdot C_{in.188}}$$ $$Z2_b(f) := \frac{1}{2j \cdot \pi \cdot f \cdot C_{i.1.tot.188.b}}$$

$$T_{i.1.b}(f) := \cfrac{Z2_b(f) \cdot \left(\cfrac{1}{R_{g1.188}} + \cfrac{1}{R_{gg1.188} + Z2_b(f)} \right)^{-1}}{\left(Z2_b(f) + R_{gg1.188} \right) \cdot \left[R_S + Z1(f) + \left(\cfrac{1}{R_{g1.188}} + \cfrac{1}{R_{gg1.188} + Z2_b(f)} \right)^{-1} \right]}$$

$$T_{i.1.b.e}(f) := 20 \cdot \log\left(\left| T_{i.1.b}(f) \right| \right) \qquad\qquad \phi_{i.1.b}(f) := atan\left(\frac{Im\left(T_{i.1.b}(f) \right)}{Re\left(T_{i.1.b}(f) \right)} \right)$$

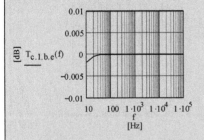

Figure 6.7 Transfer of b-i/p network   Figure 6.8 Phase of b-i/p network

$$T_{c.1.b}(f) := \left(\frac{R_{o.c1}}{R_{o.c1} + \cfrac{1}{2j \cdot \pi \cdot f \cdot C_{c1}}} \right) \qquad\qquad \phi_{c.1.b}(f) := atan\left(\frac{Im\left(T_{c.1.b}(f) \right)}{Re\left(T_{c.1.b}(f) \right)} \right)$$

$$T_{c.1.b.e}(f) := 20 \cdot \log\left(\left| T_{c.1.b}(f) \right| \right)$$

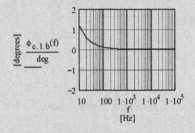

Figure 6.9 Transfer of V1 cathode network   Figure 6.10 Phase of V1 cathode network

$$Z3(f) := \frac{1}{2j \cdot \pi \cdot f \left(C_{o.1.tot.188} + C_{i.2.tot.188} \right)}$$

$$T_{o.1.b}(f) := \frac{Z3(f)}{R_{o.a1.188.b} + Z3(f)} \qquad\qquad \phi_{o.1.b}(f) := atan\left(\frac{Im\left(T_{o.1.b}(f) \right)}{Re\left(T_{o.1.b}(f) \right)} \right)$$

$$T_{o.1.b.e}(f) := 20 \cdot \log\left(\left| T_{o.1.b}(f) \right| \right) \qquad\qquad T_{o.1.b.e}(h) = -604.234 \times 10^{-9} \quad [dB]$$

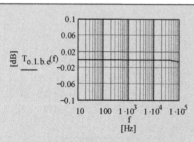

Figure 6.11 Transfer of V1 b-o/p network   Figure 6.12 Phase of V1 b-o/p network

$$Z4(f) := \frac{1}{2j \cdot \pi \cdot f \cdot C_{o.2.tot.188}}$$
$$Z5(f) := \frac{1}{2j \cdot \pi \cdot f \cdot C_{out.188}}$$

$$T_{o.2.b}(f) := \frac{\left(\dfrac{1}{Z4(f)} + \dfrac{1}{Z5(f) + R_L}\right)^{-1}}{R_{o.a2.188.b} + \left(\dfrac{1}{Z4(f)} + \dfrac{1}{Z5(f) + R_L}\right)^{-1}} \cdot \frac{R_L}{R_L + Z5(f)}$$
$$\phi_{o.2.b}(f) := atan\left(\frac{Im\left(T_{o.2.b}(f)\right)}{Re\left(T_{o.2.b}(f)\right)}\right)$$

$$T_{o.2.b.e}(f) := 20 \cdot log\left(\left|T_{o.2.b}(f)\right|\right)$$

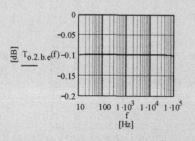
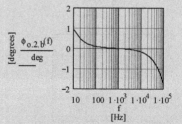

Figure 6.13 Transfer of V2 b-o/p network   Figure 6.14 Phase of V2 b-o/p network

$$T_{tot.188.b}(f) := T_{i.1.b}(f) \cdot T_{c.1.b}(f) \cdot T_{o.1.b}(f) \cdot T_{o.2.b}(f) \cdot G_{cas.188.b}$$

$$T_{tot.188.b.e}(f) := 20 \cdot log\left(\left|T_{tot.188.b}(f)\right|\right)$$
$$\phi_{G.188}(f) := -180 deg$$

$$\phi_{tot.188.b}(f) := \phi_{i.1.b}(f) + \phi_{c.1.b}(f) + \phi_{o.1.b}(f) + \phi_{o.2.b}(f) + \phi_{G.188}(f)$$

6.7.10 Gain stage frequency and phase response
- un-bypassed version:

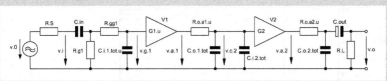

Figure 6.15 = Figure 6.3

$$Z2_u(f) := \frac{1}{2j \cdot \pi \cdot f \cdot C_{i.1.tot.188.u}}$$

$$T_{i.1.u}(f) := \frac{Z2_u(f) \cdot \left(\dfrac{1}{R_{g1.188}} + \dfrac{1}{R_{gg1.188} + Z2_u(f)} \right)^{-1}}{\left(Z2_u(f) + R_{gg1.188} \right) \cdot \left[R_S + Z1(f) + \left(\dfrac{1}{R_{g1.188}} + \dfrac{1}{R_{gg1.188} + Z2_u(f)} \right)^{-1} \right]}$$

$$\phi_{i.1.u}(f) := atan\left(\frac{Im\left(T_{i.1.u}(f) \right)}{Re\left(T_{i.1.u}(f) \right)} \right)$$

$$T_{o.1.u}(f) := \frac{Z3(f)}{R_{o.a1.188.u} + Z3(f)}$$

$$\phi_{o.1.u}(f) := atan\left(\frac{Im\left(T_{o.1.u}(f) \right)}{Re\left(T_{o.1.u}(f) \right)} \right)$$

$$T_{o.2.u}(f) := \frac{\left(\dfrac{1}{Z4(f)} + \dfrac{1}{Z5(f) + R_L} \right)^{-1}}{R_{o.a2.188.u} + \left(\dfrac{1}{Z4(f)} + \dfrac{1}{Z5(f) + R_L} \right)^{-1}} \cdot \frac{R_L}{R_L + Z5(f)}$$

$$\phi_{o.2.u}(f) := atan\left(\frac{Im\left(T_{o.2.u}(f) \right)}{Re\left(T_{o.2.u}(f) \right)} \right)$$

$$T_{o.2.u.e}(f) := 20 \cdot log\left(\left| T_{o.2.u}(f) \right| \right)$$

$$T_{tot.188.u}(f) := T_{i.1.u}(f) \, T_{o.1.u}(f) \cdot T_{o.2.u}(f) \cdot G_{cas.188.u}$$

$$T_{tot.188.u.e}(f) := 20 \cdot log\left(\left| T_{tot.188.u}(f) \right| \right) \qquad\qquad \phi_{G.188}(f) := -180 deg$$

$$\phi_{tot.188.u}(f) := \phi_{i.1.u}(f) + \phi_{o.1.u}(f) + \phi_{o.2.u}(f) + \phi_{G.188}(f)$$

6.7.11 Frequency and phase response plots:

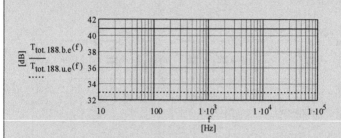

Figure 6.16 Frequency responses of the bypassed and un-bypassed CAS gain stage

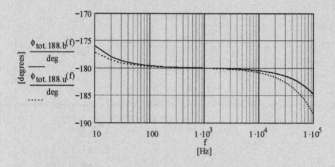

Figure 6.17 Phase responses of the bypassed and un-bypassed CAS gain stage

6.7.12 Input impedances:

$$Z_{in.b}(f) := \left[\left(R_{gg1.188} + \frac{1}{2j \cdot \pi \cdot f \cdot C_{i.1.tot.188.b}}\right)^{-1} + \frac{1}{R_{g1.188}}\right]^{-1} + \frac{1}{2j \cdot \pi \cdot f \cdot C_{in.188}}$$

$$Z_{in.u}(f) := \left[\left(R_{gg1.188} + \frac{1}{2j \cdot \pi \cdot f \cdot C_{i.1.tot.188.u}}\right)^{-1} + \frac{1}{R_{g1.188}}\right]^{-1} + \frac{1}{2j \cdot \pi \cdot f \cdot C_{in.188}}$$

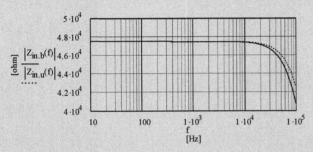

Figure 6.18 Input impedances for the bypassed and un-bypassed versions

Chapter 7 The Shunt Regulated Push-Pull Gain Stage (SRPP)

7.1   Circuit diagram

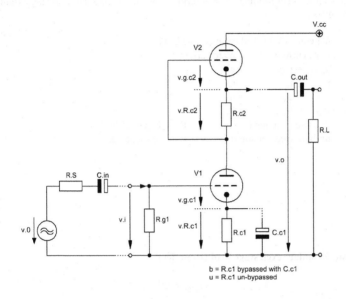

Figure 7.1 Basic design of the Shunt Regulated Push-Pull Gain Stage (SRPP)

7.2 Basic assumptions

Generally, the whole SRPP gain stage consists of a CCS or CCS+Cc gain stage (V1 with gain G1) of Chapters 1 or 2 with an active constant current source of Chapter 5.1 as the plate load (CCSo-lo with V2 and gain G2). This active load also acts like an improved cathode follower $CF2_u$ of Chapter 3.5. In this connection the impedance of the Figure 7.1 V1 plus R_{c1}/C_{c1} form the resistance $R_{c.2}$ of Figure 3.4.

To get all relevant formulae shown further down these lines the following assumption were made:

- V1 equals V2 (ideal case: double triode)
- plate current V1   = plate current V2
- $g_{m.v1}$ = $g_{m.v2}$ = $g_{m.t}$
- $r_{a.v1}$ = $r_{a.v2}$ = $r_{a.t}$
- μ_{v1} = μ_{v2} = μ_t
- R_{c2} = R_{c1} = R_c
- bypassed version (b): R_{c1} = bypassed by C_{c1}
- un-bypassed version (u): R_{c1} = un-bypassed

7.3 Basic formulae (excl. stage load (R_L)-effect)

$$G_{srpp} = G1\, G2 \tag{7.1}$$

7.3.1 bypassed version - gain $G_{srpp.b}$ in terms of $g_{m.t}$:

$$G_{srpp.b} = G1_b\, G2_b$$

$$G1_b = -g_{m.t}\, \frac{r_{a.t}\left[r_{a.t} + (1 + g_{m.t}r_{a.t})R_c\right]}{2r_{a.t} + (1 + g_{m.t}r_{a.t})R_c} \tag{7.2}$$

$$G2_b = \frac{g_{m.t}\left(\dfrac{1}{r_{a.t}} + \dfrac{1}{R_c + r_{a.t}}\right)^{-1}}{1 + g_{m.t}\left(\dfrac{1}{r_{a.t}} + \dfrac{1}{R_c + r_{a.t}}\right)^{-1}}$$

bypassed version - gain $G_{srpp.b}$ in terms of μ_t:

$$G_{srpp.b} = -\mu_t^2\, \frac{\left[r_{a.t} + (1 + \mu_t)R_c\right](R_c + r_{a.t})}{\left[2r_{a.t} + (1 + \mu_t)R_c\right]\left[2r_{a.t} + (1 + \mu_t)R_c + \mu_t r_{a.t}\right]} \tag{7.3}$$

rule of thumb for $G_{srpp.b}$:

$$G2_{b.rot} = 1$$

$$G_{srpp.b.rot} = \mu_t \frac{r_{a.t} + (1 + \mu_t) R_c}{2 r_{a.t} + (1 + \mu_t) R_c} \tag{7.4}$$

7.3.2 un-bypassed version - gain $G_{srpp.u}$ in terms of $g_{m.t}$ and μ_t:

$$G_{srpp.u} = G1_u \, G2_u$$

$$G1_u = -\frac{1}{2} g_{m.t} r_{a.t} \tag{7.5}$$

$$G2_u = \frac{g_{m.t} \left(\dfrac{1}{r_{a.t}} + \dfrac{1}{r_{a.t} + (2 + \mu_t) R_c} \right)^{-1}}{1 + g_{m.t} \left(\dfrac{1}{r_{a.t}} + \dfrac{1}{r_{a.t} + (2 + \mu_t) R_c} \right)^{-1}}$$

rule of thumb for $G_{srpp.u}$:

$$G2_{u.rot} = 1$$

$$G_{srpp.u.rot} = -\frac{1}{2} \mu_t \tag{7.6}$$

7.3.3 bypassed version - V2 cathode output resistance $R_{o.c.v2.b}$:

$$R_{o.c.v2.b} = r_{a.t} \left[\frac{r_{a.t} + R_c}{(2 + \mu_t) r_{a.t} + (1 + \mu_t) R_c} \right] \tag{7.7}$$

7.3.4 un-bypassed version - V2 cathode output resistance $R_{o.c.v2.u}$:

$$R_{o.c.v2.u} = r_{a.t} \left[\frac{r_{a.t} + (2 + \mu_t) R_c}{(2 + \mu_t) r_{a.t} + (2 + 3\mu_t + \mu_t^2) R_c} \right] \tag{7.8}$$

7.3.5 V1-R_{c1} bypass capacitance C_{c1}:

To get a flat frequency response in B_{20k} as well as a phase response deviation of $< 1°$ at 20Hz C_{c1} should be calculated as follows:

$$C_{c1} = \frac{1}{\pi f_{c.opt}} \frac{\left[r_{a.t} + (1 + \mu_t) R_c \right]}{\left[2 r_{a.t} + (1 + \mu_t) R_c \right] R_c} \tag{7.9}$$

$$f_{c.opt} = 0.2 \text{Hz} \tag{7.10}$$

7.4   Derivations

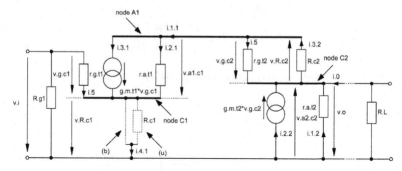

Figure 7.2   Equivalent circuit of Figure 7.1

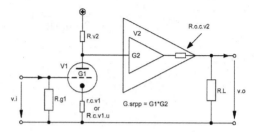

Figure 7.3   Simplified equivalent circuit of Figure 7.2

7.4.1 Gains:

Instead of producing a lot of equations derived from Figure 7.2 to get the respective gain formulae we take the much easier to handle Figure 7.3. It shows that the SRPP gain stage is made of two gain blocks G1 and G2.

$G1_b$ is a CCS+Cc gain stage with AC-grounded cathode à la Chapter 2. $G1_u$ is a CCS gain stage à la Chapter 1. G2 is an improved cathode follower $CF2_u$ à la Chapter 3.5 as well as a constant current source (CCSo-lo) à la Chapter 5.1. The impedance of the SRPP's V1 plus its components form the $CF2_u$'s lower cathode resistance R_{c2} whereas R_{c2} of the Figure 7.1 SRPP forms the $CF2_u$'s upper cathode resistance R_{c1}.

Without big extra derivation efforts the respective formulae of the chapters mentioned can be used to calculate the gains $G_{srpp.b}$ or $G_{srpp.u}$.

7.4.1.1   bypassed version:

with the upper valve's active AC resistance R_{v2} (seen from node A1):

$$R_{v2} = r_{a.t} + (1 + g_{m.t}r_{a.t})R_c \tag{7.11}$$

in terms of $g_{m.t}$ $G_{srpp.b}$ becomes:

$$Gl_b = -g_{m.t}\left(R_{v2} \parallel r_{a.t}\right) \tag{7.12}$$

$$Gl_b = -g_{m.t}\left(\frac{1}{R_{v2}} + \frac{1}{r_{a.t}}\right)^{-1} \tag{7.13}$$

$$Gl_b = -g_{m.t}\frac{r_{a.t}\left[r_{a.t} + (1 + g_{m.t}r_{a.t})R_c\right]}{2r_{a.t} + (1 + g_{m.t}r_{a.t})R_c} \tag{7.14}$$

$$G2_b = \frac{g_{m.t}\left(\dfrac{1}{r_{a.t}} + \dfrac{1}{R_c + r_{a.t}}\right)^{-1}}{1 + g_{m.t}\left(\dfrac{1}{r_{a.t}} + \dfrac{1}{R_c + r_{a.t}}\right)^{-1}} \tag{7.15}$$

$$G_{srpp.b} = Gl_b\, G2_b \tag{7.16}$$

$$G_{srpp.b} = -g_{m.t}{}^2\frac{r_{a.t}{}^2\left[r_{a.t} + (1 + g_{m.t}r_{a.t})R_c\right](R_c + r_{a.t})}{\left[2r_{a.t} + (1 + g_{m.t}r_{a.t})R_c\right]\left[2r_{a.t} + (1 + g_{m.t}r_{a.t})R_c + g_{m.t}r_{a.t}{}^2\right]} \tag{7.17}$$

since $\mu_t = g_{m.t}*r_{a.t}$ in terms of μ_t $G_{srpp.b}$ becomes:

$$G_{srpp.b} = -\mu_t{}^2\frac{\left[r_{a.t} + (1 + \mu_t)R_c\right](R_c + r_{a.t})}{\left[2r_{a.t} + (1 + \mu_t)R_c\right]\left[2r_{a.t} + (1 + \mu_t)R_c + \mu_t r_{a.t}\right]} \tag{7.18}$$

7.4.1.2   un-bypassed version:

in terms of $g_{m.t}$ $G_{srpp.u}$ becomes:

$$Gl_u = -g_{m.t}\frac{r_{a.t}\,R_{v2}}{r_{a.t} + R_{v2} + (1 + g_{m.t}\,r_{a.t})R_c} \tag{7.19}$$

$$Gl_u = -\frac{1}{2}g_{m.t}\,r_{a.t} \tag{7.20}$$

since $\mu_t = g_{m.t}*r_{a.t}$ in terms of μ_t Gl_u becomes:

$$Gl_u = -\frac{1}{2}\mu_t \tag{7.21}$$

$$G2_u = \frac{g_{m.t} \left(\dfrac{1}{r_{a.t}} + \dfrac{1}{r_{a.t} + (2 + \mu_t) R_c} \right)^{-1}}{1 + g_{m.t} \left(\dfrac{1}{r_{a.t}} + \dfrac{1}{r_{a.t} + (2 + \mu_t) R_c} \right)^{-1}} \tag{7.22}$$

$$G_{srpp.u} = G1_u \, G2_u \tag{7.23}$$

7.4.2 Specific resistances and capacitance C_{c1}:

7.4.2.1   bypassed version - V1 plate output resistances $R_{o.a.v1.b}$:

$$R_{o.a.v1.b} = r_{a.t} \tag{7.24}$$

7.4.2.2   bypassed version - V2 cathode output resistance $R_{o.c.v2.b}$:

V2 cathode resistance $r_{c.v2}$:

$$r_{c.v2} = \frac{r_{a.t}}{\mu_t + 1} \tag{7.25}$$

V2 cathode load resistance $R1_b$:

$$R1_b = r_{a.t} + R_c \tag{7.26}$$

output resistance $R_{o.c.v2.b}$:

$$R_{o.c.v2.b} = R1_b \parallel r_{c.v2}$$
$$= \left(\frac{1}{R1_b} + \frac{1}{r_{c.v2}} \right)^{-1} \tag{7.27}$$

$$R_{o.c.v2.b} = r_{a.t} \left[\frac{r_{a.t} + R_c}{r_{a.t} (2 + \mu_t) + R_c (1 + \mu_t)} \right] \tag{7.28}$$

7.4.2.3   V1 bypass capacitance C_{c1}:

sine R_{v2} is:

$$R_{v2} = r_{a.t} + (\mu_t + 1) R_c \tag{7.29}$$

V1 cathode resistance $r_{c.v1}$ becomes:

$$r_{c.v1} = \frac{r_{a.t} + R_{v2}}{1 + \mu_t} \tag{7.30}$$

V1 cathode resistance $R_{c.v1}$ becomes:

$$R_{c.v1} = r_{c.v1} \parallel R_c$$
$$= \left(\frac{1}{r_{c.v1}} + \frac{1}{R_c} \right)^{-1} \tag{7.31}$$

and capacitance C_{c1} becomes:

$$C_{c1} = \frac{1}{2\pi f_{c.opt} R_{c.v1}} \tag{7.32}$$

$$\Rightarrow C_{c1} = \frac{1}{\pi f_{c.opt}} \frac{\left[r_{a.t} + (1 + \mu_t) R_c \right]}{\left[2r_{a.t} + (1 + \mu_t) R_c \right] R_c} \tag{7.33}$$

7.4.2.4   un-bypassed version - V1 anode output resistance $R_{o.a.v1.u}$:

$$R_{o.a.v1.u} = r_{a.t} + (1 + \mu_t) R_c \tag{7.34}$$

7.4.2.5   un-bypassed version - V2 cathode output resistance $R_{o.c.v2.u}$:

V2 cathode resistance $r_{c.v2}$:

$$r_{c.v2} = \frac{r_t}{\mu_t + 1} \tag{7.35}$$

V2 cathode load resistance $R1_u$:

$$R1_u = r_{a.t} + (2 + \mu_t) R_c \tag{7.36}$$

output resistance $R_{o.c.v2.b}$:

$$R_{o.c.v2.u} = R1_u \parallel r_{c.v2}$$
$$= \left(\frac{1}{R1_u} + \frac{1}{r_{c.v2}} \right)^{-1} \tag{7.37}$$

$$R_{o.c.v2.u} = r_{a.t} \left[\frac{r_{a.t} + (2 + \mu_t) R_c}{(2 + \mu_t) r_{a.t} + (2 + 3\mu_t + \mu_t^2) R_c} \right] \tag{7.38}$$

7.5 Input capacitances $C_{i.v1}$ and $C_{i.v2}$
Output capacitances $C_{o.v1}$ and $C_{o.v2}$

data sheet figures: $C_{g.c.v1}$ = grid-cathode capacitance of V1
                             $C_{g.a.v1}$ = grid-plate capacitance of V1
                             $C_{a.c.v1}$ = plate-cathode capacitance of V1
                             $C_{g.c.v2}$ = grid-cathode capacitance of V2
                             $C_{g.a.v2}$ = grid-plate capacitance of V2
                             $C_{a.c.v2}$ = plate-cathode capacitance of V2

to be guessed: $C_{stray.1}$ = sum of stray capacitances around V1
                             $C_{stray.2}$ = sum of stray capacitances around V2

to be ignored[1]: $C_{o.c.v2}$ = cathode capacitance of V2

Note: the Miller capacitance of the input depends on V1's gain G1 only[2]!

$$C_{i.v1.b} = \left(1 + |G1_b|\right) C_{g.a.v1} \parallel C_{g.c.v1} \parallel C_{stray.1} \tag{7.39}$$

$$C_{i.v1.u} = \left(1 + |G1_u|\right) C_{g.a.v1} \parallel C_{g.c.v1} \parallel C_{stray.1} \tag{7.40}$$

$$C_{i.v2.b} = \left(1 - G2_b\right) C_{g.c.v2} \parallel C_{g.a.v2} \parallel C_{stray.2} \tag{7.41}$$

$$C_{i.v2.u} = \left(1 - G2_u\right) C_{g.c.v2} \parallel C_{g.a.v2} \parallel C_{stray.2} \tag{7.42}$$

$$C_{o.v1} = C_{a.c.v1} \parallel C_{g.a.v1} \parallel C_{g.a.v2} \tag{7.43}$$

$$C_{o.v2} = C_{a.c.v2} \tag{7.44}$$

7.6 Input impedance Z_{in}

$$Z_{in}(f) = R_{g1} \parallel C_{i.v1} \tag{7.45}$$

Inclusion of R_{gg} and C_{in} into the calculation course leads to:

$$Z_{in.b}(f) = \left[\left(R_{gg} + \frac{1}{2j\pi f\, C_{i.v1.b}}\right)^{-1} + \frac{1}{R_{g1}}\right]^{-1} + \frac{1}{2j\pi f\, C_{in}} \tag{7.46}$$

$$Z_{in.u}(f) = \left[\left(R_{gg} + \frac{1}{2j\pi f\, C_{i.v1.u}}\right)^{-1} + \frac{1}{R_{g1}}\right]^{-1} + \frac{1}{2j\pi f\, C_{in}} \tag{7.47}$$

---

[1]   to be ignored in B_{20k} only because - compared with the size of the corresponding capacitances - the output impedance is rather low

[2]   Miller capacitance effects on C_i and C_o: see Chapter 1.3.2.5

7.7 Gain stage frequency and phase response calculations

7.7.1 Bypassed version:

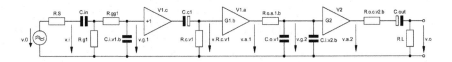

Figure 7.4 Simplified[3] equivalent circuit of Figure 7.1 (bypassed version)
- including all frequency and phase response relevant components

7.7.1.1   1st stage (V1 = V1$_c$+V1$_a$) input transfer function $T_{i.1.b}$(f) and phase $\varphi_{i.1.b}$ (f) -
including source resistance R$_S$ and an oscillation preventing resistor R$_{gg}$<<R$_g$:

$$T_{i.1.b}(f) = \frac{v_{g.1}}{v_0} \qquad (7.48)$$

$$\varphi_{i.1.b}(f) = \arctan\left\{ \frac{\text{Im}\left[T_{i.1.b}(f)\right]}{\text{Re}\left[T_{i.1.b}(f)\right]} \right\} \qquad (7.49)$$

$$T_{i.1.b}(f) = \frac{Z2_b(f)\left(\dfrac{1}{R_{g1}} + \dfrac{1}{R_{gg1} + Z2_b(f)}\right)^{-1}}{\left(R_{gg1} + Z2_b(f)\right)\left[R_S + Z1(f) + \left(\dfrac{1}{R_{g1}} + \dfrac{1}{R_{gg1} + Z2_b(f)}\right)^{-1}\right]} \qquad (7.50)$$

$$Z1(f) = \left(2j\pi f\, C_{in}\right)^{-1}$$
$$Z2_b(f) = \left(2j\pi f\, C_{i.v1.b}\right)^{-1} \qquad (7.51)$$

7.7.1.2   1st stage V1$_c$ cathode transfer function $T_{c.1.b}$(f) and phase $\varphi_{c.1.b}$(f):

$$\varphi_{c.1.b}(f) = \arctan\left\{ \frac{\text{Im}\left[T_{c.1.b}(f)\right]}{\text{Re}\left[T_{c.1.b}(f)\right]} \right\} \qquad (7.52)$$

$$T_{c.1.b}(f) = \frac{R_{c.v1}}{R_{c.v1} + \dfrac{1}{2j\pi f\, C_{c.1}}} \qquad (7.53)$$

---

3   See also footnotes 2 and 3 of Chapter 1 and footnote 3 of Chapter 2

7.7.1.3 1st stage output transfer function $T_{o.1.b}(f)$ and phase $\varphi_{o.1.b}(f)$:

$$T_{o.1.b}(f) = \frac{v_{g.2}}{v_{a.1}} \tag{7.54}$$

$$\varphi_{o.1.b}(f) = \arctan\left\{ \frac{\text{Im}\left[T_{o.1.b}(f)\right]}{\text{Re}\left[T_{o.1.b}(f)\right]} \right\} \tag{7.55}$$

$$T_{o.1.b}(f) = \frac{Z3_b(f)}{R_{o.a.1.b} + Z3_b(f)} \tag{7.56}$$

$$Z3_b(f) = \left[2j\pi f \left(C_{o.v1} + C_{i.v2.b} \right) \right]^{-1} \tag{7.57}$$

7.7.1.4 2nd stage output transfer function $T_{o.2.b}(f)$ and phase $\varphi_{o.2.b}(f)$:

$$T_{o.2.b}(f) = \frac{v_o}{v_{c.2}} \tag{7.58}$$

$$\varphi_{o.2.b}(f) = \arctan\left\{ \frac{\text{Im}\left[T_{o.2.b}(f)\right]}{\text{Re}\left[T_{o.2.b}(f)\right]} \right\} \tag{7.59}$$

$$T_{o.2.b}(f) = \frac{R_L}{R_{o.c.v2.b} + R_L + Z4(f)} \tag{7.60}$$

$$Z4(f) = \left(2j\pi f\, C_{out} \right)^{-1} \tag{7.61}$$

7.7.1.5 1st stage fundamental phase shift $\varphi_{G.v1}(f)$:

$$\varphi_{G.v1}(f) = \varphi_{G.srpp.b}(f) = -180° \tag{7.62}$$

7.7.1.6   total gain stage transfer function $T_{tot.b}(f)$ and phase $\varphi_{tot.b}(f)$:

$$T_{tot.b}(f) = T_{i.1.b}(f)\, T_{c.1.b}(f)\, T_{o.1.b}(f)\, T_{o.2.b}(f)\, G1_b\, G2_b \tag{7.63}$$

$$\varphi_{tot.b}(f) = \varphi_{i.1.b}(f) + \varphi_{c.1.b}(f) + \varphi_{o.1.b}(f) + \varphi_{o.2.b}(f) + \varphi_{G.srpp.b}(f) \tag{7.64}$$

7.7.2 Un-bypassed version:

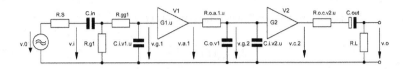

Figure 7.5 Simplified[4] equivalent circuit of Figure 7.1 (un-bypassed version)
- including all frequency and phase response relevant components

7.7.2.1   1st stage (V1) input transfer function $T_{i.1.u}(f)$ and phase $\varphi_{i.1.u}(f)$ - including source resistance R_S and an oscillation preventing resistor $R_{gg} \ll R_g$:

$$T_{i.1.u}(f) = \frac{v_{g.1}}{v_0} \tag{7.65}$$

$$\varphi_{i.1.u}(f) = \arctan\left\{\frac{Im\left[T_{i.1.u}(f)\right]}{Re\left[T_{i.1.u}(f)\right]}\right\} \tag{7.66}$$

$$T_{i.1.u}(f) = \frac{Z2_u(f)\left(\dfrac{1}{R_{g1}} + \dfrac{1}{R_{gg1} + Z2_u(f)}\right)^{-1}}{\left(R_{gg1} + Z2_u(f)\right)\left[R_S + Z1(f) + \left(\dfrac{1}{R_{g1}} + \dfrac{1}{R_{gg1} + Z2_u(f)}\right)^{-1}\right]} \tag{7.67}$$

$$Z1(f) = \left(2j\pi f\, C_{in}\right)^{-1}$$
$$Z2_u(f) = \left(2j\pi f\, C_{i.v1.u}\right)^{-1} \tag{7.68}$$

7.7.2.2   1st stage output transfer function $T_{o.1.u}(f)$ and phase $\varphi_{o.1.u}(f)$:

$$T_{o.1.u}(f) = \frac{v_{g.2}}{v_{a.1}} \tag{7.69}$$

$$\varphi_{o.1.u}(f) = \arctan\left\{\frac{Im\left[T_{o.1.u}(f)\right]}{Re\left[T_{o.1.u}(f)\right]}\right\} \tag{7.70}$$

$$T_{o.1.u}(f) = \frac{Z3_u(f)}{R_{o.a.1.b} + Z3_u(f)} \tag{7.71}$$

---

[4]   See also footnotes 2 and 3 of Chapter 1

$$Z3_u(f) = \left[2j\pi f \left(C_{o.v1} + C_{i.v2.u}\right)\right]^{-1} \tag{7.72}$$

7.7.2.3 2nd stage output transfer function $T_{o.2.u}(f)$ and phase $\varphi_{o.2.u}(f)$:

$$T_{o.2.u}(f) = \frac{v_o}{v_{c.2}} \tag{7.73}$$

$$\varphi_{o.2.u}(f) = \arctan\left\{\frac{Im\left[T_{o.2.u}(f)\right]}{Re\left[T_{o.2.u}(f)\right]}\right\} \tag{7.74}$$

$$T_{o.2.u}(f) = \frac{R_L}{R_{o.c.v2.u} + R_L + Z4(f)} \tag{7.75}$$

$$Z4(f) = \left(2j\pi f \, C_{out}\right)^{-1} \tag{7.76}$$

7.7.2.4 1st stage fundamental phase shift $\varphi_{G.v1}(f)$:

$$\varphi_{G.v1}(f) = \varphi_{G.srpp.u}(f) = -180° \tag{7.77}$$

7.7.2.5   total gain stage transfer function $T_{tot.u}(f)$ and phase $\varphi_{tot.u}(f)$:

$$T_{tot.u}(f) = T_{i.1.u}(f) T_{o.1.u}(f) T_{o.2.u}(f) G1_u G2_u \tag{7.78}$$

$$\varphi_{tot.u}(f) = \varphi_{i.1.u}(f) + \varphi_{o.1.u}(f) + \varphi_{o.2.u}(f) + \varphi_{G.srpp.u}(f) \tag{7.79}$$

7.8   Example with ECC83 / 12AX7 (83):

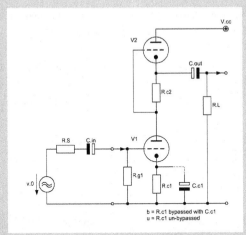

Figure 7.6 SRPP example circuitry

7.8.1   Triode bias data:

$I_{a1.83} = I_{a2.83} = 1.2 \cdot 10^{-3} A$ $\qquad V_{a1.83} = 100V \qquad\qquad V_{a2.83} = 100V$

$V_{cc} = 201V \qquad\qquad\qquad V_{g1.83} = -0.5V \qquad\qquad V_{g2.83} = 100.5V$

7.8.2   Triode valve constants:

$g_{m.83} := 1.75 \cdot 10^{-3} \cdot S \qquad\quad \mu_{83} := 101.5 \qquad\qquad r_{a.83} := 58 \cdot 10^{3} \Omega$

$g_{m.v1} := g_{m.83} \qquad\qquad\quad \mu_{v1} := \mu_{83} \qquad\qquad r_{a.v1} := r_{a.83}$

$g_{m.v2} := g_{m.83} \qquad\qquad\quad \mu_{v2} := \mu_{83} \qquad\qquad r_{a.v2} := r_{a.83}$

$C_{g.c.1.83} := 1.65 \cdot 10^{-12} F \qquad C_{g.a.1.83} := 1.6 \cdot 10^{-12} F \qquad C_{a.c.1.83} := 0.33 \cdot 10^{-12} F$

$C_{g.c.2.83} := 1.65 \cdot 10^{-12} F \qquad C_{g.a.2.83} := 1.6 \cdot 10^{-12} F \qquad C_{a.c.2.83} := 0.33 \cdot 10^{-12} F$

7.8.3   Circuit variables:

$R_{c1.83} := 417\Omega \qquad\qquad R_{c2.83} := R_{c1.83} \qquad\qquad R_{c.83} := R_{c1.83}$

$R_{S} := 1 \cdot 10^{3} \Omega \qquad\qquad R_{L} := 47.5 \cdot 10^{3} \Omega \qquad\quad R_{gg1.83} := 0.22 \cdot 10^{3} \Omega$

$R_{g1.83} := 47.5 \cdot 10^{3} \Omega \qquad C_{in.83} := 2.2 \cdot 10^{-6} F \qquad\quad C_{out.83} := 10 \cdot 10^{-6} F$

$\qquad\qquad\qquad\qquad\quad C_{stray.83.1} := 10 \cdot 10^{-12} F \qquad C_{stray.83.2} := 10 \cdot 10^{-12} F$

➢   MCD Worksheet VII   SRPP calculations   page 2

7.8.4   Calculation relevant data:

frequency range f for the below shown graphs: $f := 10\,Hz, 20\,Hz .. 100000\,Hz$

$h := 1000 \cdot Hz$

7.8.5   Gain $G_{srpp.83.b}$:

$R_{v2} := r_{a.v2} + \left(1 + \mu_{v2}\right) \cdot R_{c2.83}$ $R_{v2} = 100.743 \times 10^{3}\,\Omega$

$G1_b := -g_{m.83} \cdot \left(\dfrac{1}{R_{v2}} + \dfrac{1}{r_{a.83}} \right)^{-1}$ via V1 $G1_b = -64.415 \times 10^{0}$

$G2_b := \dfrac{g_{m.v2} \left(\dfrac{1}{r_{a.v2}} + \dfrac{1}{R_{c2.83} + r_{a.vl}} \right)^{-1}}{1 + g_{m.v2} \left(\dfrac{1}{r_{a.v2}} + \dfrac{1}{R_{c2.83} + r_{a.vl}} \right)^{-1}}$ via V2 $G2_b = 980.744 \times 10^{-3}$

$G_{srpp.83.b} := G1_b \cdot G2_b$ $G_{srpp.83.b} = -63.174 \times 10^{0}$

$G_{srpp.83.b.e} := 20 \cdot \log\left(\left| G_{srpp.83.b} \right| \right)$ $G_{srpp.83.b.e} = 36.011 \times 10^{0}$ [dB]

$G_{srpp.83.b.rot} := G1_b$ $G_{srpp.83.b.rot} = -64.415 \times 10^{0}$

$G_{srpp.83.b.rot.e} := 20 \cdot \log\left(\left| G_{srpp.83.b.rot} \right| \right)$ $G_{srpp.83.b.rot.e} = 36.18 \times 10^{0}$ [dB]

$$G_{srpp.83.b.e} - G_{srpp.83.b.rot.e} = -168.887 \times 10^{-3}$$

7.8.6   Gain $G_{srpp.83.u}$:

$G1_u := -\dfrac{1}{2} \cdot g_{m.83} \cdot r_{a.83}$ $G1_u = -50.75 \times 10^{0}$

$G2_u := \dfrac{g_{m.v2} \left[\dfrac{1}{r_{a.v2}} + \dfrac{1}{R_{c2.83} + r_{a.vl} + \left(1 + \mu_{vl}\right) \cdot R_{c1.83}} \right]^{-1}}{1 + g_{m.v2} \left[\dfrac{1}{r_{a.v2}} + \dfrac{1}{R_{c2.83} + r_{a.vl} + \left(1 + \mu_{vl}\right) \cdot R_{c1.83}} \right]^{-1}}$ $G2_u = 984.736 \times 10^{-3}$

$G_{srpp.83.u} := G1_u \cdot G2_u$ $G_{srpp.83.u} = -49.975 \times 10^{0}$

$G_{srpp.83.u.e} := 20 \cdot \log\left(\left| G_{srpp.83.u} \right| \right)$ $G_{srpp.83.u.e} = 33.975 \times 10^{0}$ [dB]

$G_{srpp.83.u.rot} := G1_u$ $G_{srpp.83.u.rot} = -50.75 \times 10^{0}$

$G_{srpp.83.u.rot.e} := 20 \cdot \log\left(\left| G_{srpp.83.u.rot} \right| \right)$ $G_{srpp.83.u.rot.e} = 34.109 \times 10^{0}$ [dB]

$$G_{srpp.83.u.e} - G_{srpp.83.u.rot.e} = -133.607 \times 10^{-3}$$

7.8.7 Specific resistances:

$$R_{o.a.v1.b} := r_{a.v1}$$

$$R_{o.a.v1.b} = 58 \times 10^3 \, \Omega$$

$$R_{o.a.v1.u} := r_{a.v1} + \left(1 + \mu_{v1}\right) \cdot R_{c1.83}$$

$$R_{o.a.v1.u} = 100.743 \times 10^3 \, \Omega$$

$$r_{c.v1} := \frac{r_{a.v1} + R_{v2}}{1 + \mu_{v1}}$$

$$r_{c.v1} = 1.549 \times 10^3 \, \Omega$$

$$R_{c.v1} := \left(\frac{1}{r_{c.v1}} + \frac{1}{R_{c1.83}}\right)^{-1}$$

$$R_{c.v1} = 328.539 \times 10^0 \, \Omega$$

$$r_{c.v2} := \frac{r_{a.v2}}{1 + \mu_{v2}}$$

$$r_{c.v2} = 565.854 \times 10^0 \, \Omega$$

$$Rl_b := r_{a.v1} + R_{c2.83}$$

$$Rl_b = 58.417 \times 10^3 \, \Omega$$

$$R_{o.c.v2.b} := \left(\frac{1}{Rl_b} + \frac{1}{r_{c.v2}}\right)^{-1}$$

$$R_{o.c.v2.b} = 560.425 \times 10^0 \, \Omega$$

$$Rl_u := r_{a.v1} + \left(1 + \mu_{v1}\right) \cdot R_{c1.83} + R_{c2.83}$$

$$Rl_u = 101.159 \times 10^3 \, \Omega$$

$$R_{o.c.v2.u} := \left(\frac{1}{Rl_u} + \frac{1}{r_{c.v2}}\right)^{-1}$$

$$R_{o.c.v2.u} = 562.706 \times 10^0 \, \Omega$$

7.8.8 Capacitance C_{c1}:

$$f_c := 20 \text{Hz} \qquad\qquad f_{c.opt} := \frac{f_c}{100}$$

$$f_{c.opt} = 200 \times 10^{-3} \, \text{Hz}$$

$$C_{c1} := \frac{1}{2 \cdot \pi \cdot f_{c.opt} \cdot R_{c.v1}}$$

$$C_{c1} = 2.422 \times 10^{-3} \, \text{F}$$

7.8.9 Specific input and output capacitances:

$$C_{i.v1.b} := \left(1 + |Gl_b|\right) \cdot C_{g.a.1.83} + C_{g.c.1.83} + C_{stray.83.1}$$

$$C_{i.v1.b} = 116.314 \times 10^{-12} \, \text{F}$$

$$C_{i.v1.u} := \left(1 + |Gl_u|\right) \cdot C_{g.a.1.83} + C_{g.c.1.83} + C_{stray.83.1}$$

$$C_{i.v1.u} = 94.45 \times 10^{-12} \, \text{F}$$

$$C_{i.v2.b} := \left(1 - G2_b\right) \cdot C_{g.a.2.83} + C_{g.c.2.83} + C_{stray.83.2}$$

$$C_{i.v2.b} = 11.681 \times 10^{-12} \, \text{F}$$

$$C_{i.v2.u} := \left(1 - G2_u\right) \cdot C_{g.a.2.83} + C_{g.c.2.83} + C_{stray.83.2}$$

$$C_{i.v2.u} = 11.674 \times 10^{-12} \, \text{F}$$

$$C_{o.v1} := C_{a.c.1.83} + C_{g.a.1.83} + C_{g.a.2.83}$$

$$C_{o.v2} := C_{a.c.2.83}$$

7.8.10 Gain stage frequency and phase response
- bypassed version:

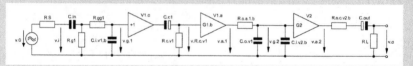

Figure 7.7 = Figure 7.4

$$Z1(f) := \frac{1}{2j \cdot \pi \cdot f \cdot C_{in.83}}$$

$$Z2_b(f) := \frac{1}{2j \cdot \pi \cdot f \cdot C_{i.v1.b}}$$

$$T_{i.1.b}(f) := \frac{Z2_b(f) \cdot \left(\dfrac{1}{R_{g1.83}} + \dfrac{1}{R_{gg1.83} + Z2_b(f)} \right)^{-1}}{\left(Z2_b(f) + R_{gg1.83} \right) \cdot \left[R_S + Z1(f) + \left(\dfrac{1}{R_{g1.83}} + \dfrac{1}{R_{gg1.83} + Z2_b(f)} \right)^{-1} \right]}$$

$$T_{i.1.b.e}(f) := 20 \cdot \log\left(\left| T_{i.1.b}(f) \right| \right)$$

$$\phi_{i.1.b}(f) := atan\left(\frac{Im\left(T_{i.1.b}(f) \right)}{Re\left(T_{i.1.b}(f) \right)} \right)$$

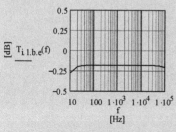

Figure 7.8 Transfer of i/p network

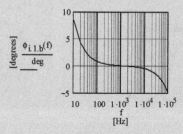

Figure 7.9 Phase of i/p network

$$T_{c.1.b}(f) := \left(\frac{R_{c.v1}}{R_{c.v1} + \dfrac{1}{2j \cdot \pi \cdot f \cdot C_{c1}}} \right)$$

$$\phi_{c.1.b}(f) := atan\left(\frac{Im\left(T_{c.1.b}(f) \right)}{Re\left(T_{c.1.b}(f) \right)} \right)$$

$$T_{c.1.b.e}(f) := 20 \cdot \log\left(\left| T_{c.1.b}(f) \right| \right)$$

➤ MCD Worksheet VII   SRPP calculations   page 5

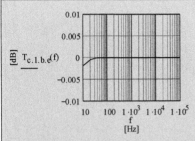

Figure 7.10 Transfer of V1 cathode network

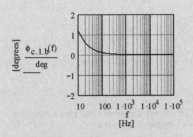

Figure 7.11 Phase of V1 cathode network

$$Z3_b(f) := \frac{1}{2j \cdot \pi \cdot f \cdot \left(C_{o.v1} + C_{i.v2.b}\right)}$$

$$T_{o.1.b}(f) := \frac{Z3_b(f)}{R_{o.a.v1.b} + Z3_b(f)}$$

$$\phi_{o.1.b}(f) := atan\left(\frac{Im\left(T_{o.1.b}(f)\right)}{Re\left(T_{o.1.b}(f)\right)}\right)$$

$$T_{o.1.b.e}(f) := 20 \cdot log\left(\left|T_{o.1.b}(f)\right|\right)$$

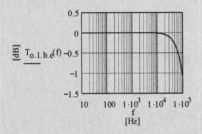

Figure 7.12 Transfer of V1 o/p network

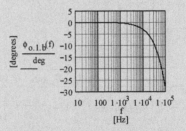

Figure 7.13 Phase of V1 o/p network

$$Z4(f) := \frac{1}{2j \cdot \pi \cdot f \cdot C_{out.83}}$$

$$T_{o.2.b}(f) := \frac{R_L}{R_{o.c.v2.b} + R_L + Z4(f)}$$

$$T_{o.2.b.e}(f) := 20 \cdot log\left(\left|T_{o.2.b}(f)\right|\right)$$

$$\phi_{o.2.b}(f) := atan\left(\frac{Im\left(T_{o.2.b}(f)\right)}{Re\left(T_{o.2.b}(f)\right)}\right)$$

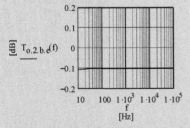

Figure 7.14 Transfer of V2 o/p network

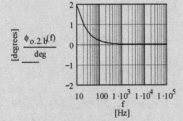

Figure 7.15 Phase of V2 o/p network

➢ **MCD Worksheet VII**   **SRPP calculations**   page 6

$$T_{tot.83.b}(f) := T_{i.1.b}(f) \cdot T_{c.1.b}(f) \, T_{o.1.b}(f) \cdot T_{o.2.b}(f) \cdot G1_b \cdot G2_b$$

$$T_{tot.83.b.e}(f) := 20 \cdot \log\left(\left|T_{tot.83.b}(f)\right|\right)$$ $$\phi_{G.83}(f) := -180 \deg$$

$$\phi_{tot.83.b}(f) := \phi_{i.1.b}(f) + \phi_{c.1.b}(f) + \phi_{o.1.b}(f) + \phi_{o.2.b}(f) + \phi_{G.83}(f)$$

7.8.11 Gain stage frequency and phase response
       - un-bypassed version:

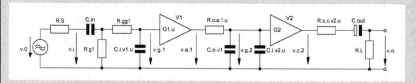

Figure 7.16 = Figure 7.5

$$Z2_u(f) := \frac{1}{2j \cdot \pi \cdot f \cdot C_{i.v1.u}}$$

$$T_{i.1.u}(f) := \frac{Z2_u(f) \cdot \left(\dfrac{1}{R_{g1.83}} + \dfrac{1}{R_{gg1.83} + Z2_u(f)}\right)^{-1}}{\left(Z2_u(f) + R_{gg1.83}\right)\left[R_S + Z1(f) + \left(\dfrac{1}{R_{g1.83}} + \dfrac{1}{R_{gg1.83} + Z2_u(f)}\right)^{-1}\right]}$$

$$Z3_u(f) := \frac{1}{2j \cdot \pi \cdot f \cdot \left(C_{o.v1} + C_{i.v2.u}\right)}$$ $$\phi_{i.1.u}(f) := \operatorname{atan}\left(\frac{\operatorname{Im}\left(T_{i.1.u}(f)\right)}{\operatorname{Re}\left(T_{i.1.u}(f)\right)}\right)$$

$$T_{o.1.u}(f) := \frac{Z3_u(f)}{R_{o.a.v1.u} + Z3_u(f)}$$ $$\phi_{o.1.u}(f) := \operatorname{atan}\left(\frac{\operatorname{Im}\left(T_{o.1.u}(f)\right)}{\operatorname{Re}\left(T_{o.1.u}(f)\right)}\right)$$

$$T_{o.2.u}(f) := \frac{R_L}{R_{o.c.v2.u} + R_L + Z4(f)}$$ $$\phi_{o.2.u}(f) := \operatorname{atan}\left(\frac{\operatorname{Im}\left(T_{o.2.u}(f)\right)}{\operatorname{Re}\left(T_{o.2.u}(f)\right)}\right)$$

$$T_{tot.83.u}(f) := T_{i.1.u}(f) \, T_{o.1.u}(f) \cdot T_{o.2.u}(f) \cdot G1_u \cdot G2_u$$

$$T_{tot.83.u.e}(f) := 20 \cdot \log\left(\left|T_{tot.83.u}(f)\right|\right)$$ $$\phi_{G.83}(f) := -180 \deg$$

$$\phi_{tot.83.u}(f) := \phi_{i.1.u}(f) + \phi_{o.1.u}(f) + \phi_{o.2.u}(f) + \phi_{G.83}(f)$$

7.8.12 Frequency and phase response plots:

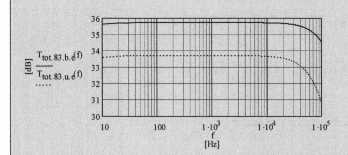

Figure 7.17 Frequency responses of the bypassed and un-bypassed
SRPP gain stage

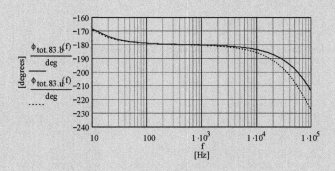

Figure 7.18 Phase responses of the bypassed and un-bypassed
SRPP gain stage

➢   MCD Worksheet VII   SRPP calculations   page 8

7.8.13 Input impedances:

$$Z_{in.b}(f) := \left[\left(R_{gg1.83} + \frac{1}{2j \cdot \pi \cdot f \cdot C_{i.vl.b}} \right)^{-1} + \frac{1}{R_{g1.83}} \right]^{-1} + \frac{1}{2j \cdot \pi \cdot f \cdot C_{in.83}}$$

$$Z_{in.u}(f) := \left[\left(R_{gg1.83} + \frac{1}{2j \cdot \pi \cdot f \cdot C_{i.vl.u}} \right)^{-1} + \frac{1}{R_{g1.83}} \right]^{-1} + \frac{1}{2j \cdot \pi \cdot f \cdot C_{in.83}}$$

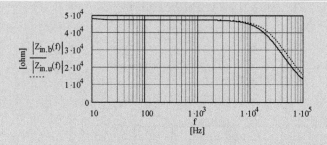

Figure 7.19 Input impedances for the bypassed and un-bypassed versions

Chapter 8 The μ-Follower Gain Stage (μ-F)

8.1 Circuit diagram

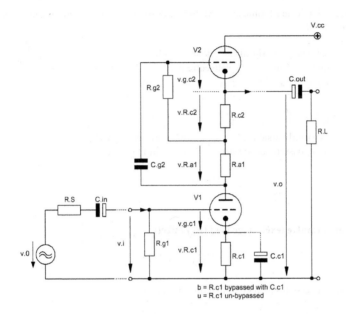

Figure 8.1 Basic design of the μ-Follower Gain Stage

8.2   Basic assumptions

Generally, the whole μ-F gain stage consists of a CCS or CCS+Cc gain stage (V1 with gain G1) of Chapters 1 or 2 with an active constant current source (CCSo-hi with V2 and gain G2) of Chapter 5.3 as the plate load. This active load also acts like an improved cathode follower $CF2_u$ of Chapter 3.5. In this connection the impedance of the Figure 8.1 V1 plus R_{c1}/C_{c1} form the resistance $R_{c.2}$ of Figure 3.4. The output voltage of V1 is fed to the input (grid) of V2 via C_{g2}.

To get all relevant formulae shown further down these lines the following assumption were made:

- V1 equals V2 (ideal case: double triode)
- plate current V1   = plate current V2
- $g_{m.v1}$ $= g_{m.v2}$ $= g_{m.t}$
- $r_{a.v1}$ $= r_{a.v2}$ $= r_{a.t}$
- μ_{v1} $= \mu_{v2}$ $= \mu_t$
- R_{c1} $= R_{c2}$ $= R_c$
- bypassed version (b): R_{c1} = bypassed by C_{c1}
- un-bypassed version (u): R_{c1} = un-bypassed

8.3   Basic formulae (excl. stage load (R_L)-effect)

$$G_{\mu.F} = G1\,G2 \tag{8.1}$$

8.3.1 bypassed version - gain $G_{\mu.F.b}$ in terms of $g_{m.t}$:

$$G_{\mu.F.b} = G1_b\,G2_b$$

$$G1_b = -g_{m.t}\frac{r_{a.t}\left[r_{a.t}+\left(1+g_{m.t}r_{a.t}\right)\left(R_c+R_{a1}\parallel R_{g2}\right)\right]}{2r_{a.t}+\left(1+g_{m.t}r_{a.t}\right)\left(R_c+R_{a1}\parallel R_{g2}\right)} \tag{8.2}$$

$$G2_b = \frac{g_{m.t}\left(\dfrac{1}{r_{a.t}}+\dfrac{1}{R_c+R_{a1}+r_{a.t}}\right)^{-1}}{1+g_{m.t}\left(\dfrac{1}{r_{a.t}}+\dfrac{1}{R_c+R_{a1}+r_{a.t}}\right)^{-1}}$$

bypassed version - gain $G_{\mu.F.b}$ in terms of μ_t:

$$G_{\mu.F.b} = -\mu_t^2\frac{\left[r_{a.t}+\left(1+\mu_t\right)\left(R_c+R_{a1}\parallel R_{g2}\right)\right]}{\left[2r_{a.t}+\left(1+\mu_t\right)\left(R_c+R_{a1}\parallel R_{g2}\right)\right]}$$
$$*\frac{\left(R_c+R_{a1}+r_{a.t}\right)}{\left[2r_{a.t}+R_c+R_{a1}+\mu_t\left(R_c+R_{a1}+r_{a.t}\right)\right]} \tag{8.3}$$

rule of thumb for $G_{\mu.F.b}$:

$$G2_{b.rot} = 1$$
$$G_{\mu.F.b.rot} = \mu_t$$

(8.4)

8.3.2 un-bypassed version - gain $G_{\mu.F.u}$ in terms of $g_{m.t}$ and μ_t:

$$G_{\mu.F.u} = G1_u\, G2_u$$

$$G1_u = -\mu_t\, \frac{r_{a.t} + (1+\mu_t)\left(R_c + R_{a1} \parallel R_{g2}\right)}{2r_{a.t} + (1+\mu_t)\left(2R_c + R_{a1} \parallel R_{g2}\right)}$$

$$G2_u = \frac{g_{m.t}\left(\dfrac{1}{r_{a.t}} + \dfrac{1}{R_{a1} + r_{a.t} + (2+\mu_t)R_c}\right)^{-1}}{1 + g_{m.t}\left(\dfrac{1}{r_{a.t}} + \dfrac{1}{R_{a1} + r_{a.t} + (2+\mu_t)R_c}\right)^{-1}}$$

(8.5)

rule of thumb for $G_{\mu.F.u}$:

$$G2_{u.rot} = 1$$
$$G_{\mu.F.u.rot} = \mu_t$$

(8.6)

8.3.3 bypassed version - V2 cathode output resistance $R_{o.c.v2.b}$:

$$R_{o.c.v2.b} = r_{a.t}\left[\frac{r_{a.t} + R_c}{(2+\mu_t)r_{a.t} + (1+\mu_t)R_c}\right]$$

(8.7)

8.3.4 un-bypassed version - V2 cathode output resistance $R_{o.c.v2.u}$:

$$R_{o.c.v2.u} = r_{a.t}\left[\frac{r_{a.t} + (2+\mu_t)R_c}{(2+\mu_t)r_{a.t} + \left(2 + 3\mu_t + \mu_t^2\right)R_c}\right]$$

(8.8)

8.3.5 V1-R_{c1} bypass capacitance C_{c1}:

To get a flat frequency response in B_{20k} as well as a phase response deviation of < 1° at 20Hz C_{c1} should be calculated as follows:

$$C_{c1} = \frac{1}{\pi f_{c.opt}}\, \frac{\left[r_{a.t} + (1+\mu_t)R_c\right]}{\left[2r_{a.t} + (1+\mu_t)R_c\right]R_c}$$

(8.9)

$$f_{c.opt} = 0.2Hz$$

(8.10)

8.4   Derivations

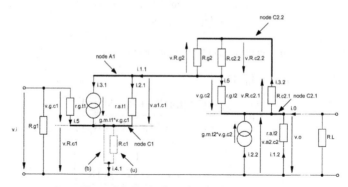

Figure 8.2 Equivalent circuit of Figure 8.1

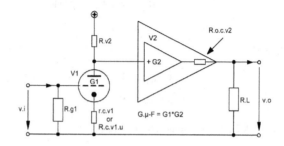

Figure 8.3 Simplified equivalent circuit of Figure 8.2

8.4.1 Gains:

Instead of producing a lot of equations derived from Figure 8.2 to get the respective gain formulae we take the much easier to handle Figure 8.3. It shows that the μ-F gain stage is made of two gain blocks G1 and G2.

G1$_b$ is a CCS+Cc gain stage with AC-grounded cathode à la Chapter 2. G1$_u$ is a CCS gain stage à la Chapter 1. G2 is an improved cathode follower CF2$_u$ à la Chapter 3.5 as well as a constant current source CCSo-hi à la Chapter 5.3. The impedance of the μ-F's V1 plus its components form the CF2$_u$'s lower cathode resistance R$_{c2}$ whereas R$_{a1}$ plus R$_{c2}$ of the Figure 8.1 μ-F form the CF2$_u$'s upper cathode resistance R$_{c1}$.

Without big extra derivation efforts the respective formulae of the chapters mentioned can be used to calculate the gains G$_{\mu.F.b}$ or G$_{\mu.F.u}$.

8.4.1.1   bypassed version:

with the upper valve's active AC resistance R$_{v2}$ (seen from node A1):

$$R_{v2} = r_{a.t} + (\mu_t + 1)(R_{c2} + R_{a1} \parallel R_{g2})$$ (8.11)

in terms of $g_{m.t}$ $G_{\mu.F.b}$ becomes:

$$G1_b = -g_{m.t}\left(R_{v2} \parallel r_{a.t}\right) \tag{8.12}$$

$$G1_b = -g_{m.t}\left(\frac{1}{R_{v2}} + \frac{1}{r_{a.t}}\right)^{-1} \tag{8.13}$$

$$G2_b = \frac{g_{m.t}\left(\dfrac{1}{r_{a.t}} + \dfrac{1}{R_c + R_{a1} + r_{a.t}}\right)^{-1}}{1 + g_{m.t}\left(\dfrac{1}{r_{a.t}} + \dfrac{1}{R_c + R_{a1} + r_{a.t}}\right)^{-1}} \tag{8.14}$$

$$G_{\mu.F.b} = G1_b\, G2_b \tag{8.15}$$

$$G_{\mu.F.b} = -g_{m.t}^2\, \frac{r_{a.t}^2\left[r_{a.t} + \left(1 + g_{m.t}r_{a.t}\right)\left(R_c + R_{a1} \parallel R_{g2}\right)\right]}{\left[2r_{a.t} + \left(1 + g_{m.t}r_{a.t}\right)\left(R_c + R_{a1} \parallel R_{g2}\right)\right]}$$
$$* \frac{\left(R_c + R_{a1} + r_{a.t}\right)}{\left[2r_{a.t} + R_c + R_{a1} + g_{m.t}r_{a.t}\left(R_c + R_{a1} + r_{a.t}\right)\right]} \tag{8.16}$$

since $\mu_t = g_{m.t}*r_{a.t}$ in terms of μ_t $G_{\mu.F.b}$ becomes:

$$G_{\mu.F.b} = -\mu_t^2\, \frac{\left[r_{a.t} + \left(1 + \mu_t\right)\left(R_c + R_{a1} \parallel R_{g2}\right)\right]}{\left[2r_{a.t} + \left(1 + \mu_t\right)\left(R_c + R_{a1} \parallel R_{g2}\right)\right]}$$
$$* \frac{\left(R_c + R_{a1} + r_{a.t}\right)}{\left[2r_{a.t} + R_c + R_{a1} + \mu_t\left(R_c + R_{a1} + r_{a.t}\right)\right]} \tag{8.17}$$

8.4.1.2   un-bypassed version:

in terms of $g_{m.t}$ $G_{\mu.F.u}$ becomes:

$$G1_u = -g_{m.t}\, \frac{r_{a.t}\, R_{v2}}{r_{a.t} + R_{v2} + R_c(1 + g_{m.t}\, r_{a.t})} \tag{8.18}$$

since $\mu_t = g_{m.t}*r_{a.t}$ in terms of μ_t $G1_u$ becomes:

$$G1_u = -\mu_t\, \frac{r_{a.t} + \left(1 + \mu_t\right)\left(R_c + R_{a1} \parallel R_{g2}\right)}{2r_{a.t} + \left(1 + \mu_t\right)\left(2R_c + R_{a1} \parallel R_{g2}\right)} \tag{8.19}$$

$$G2_u = \frac{g_{m.t}\left(\dfrac{1}{r_{a.t}} + \dfrac{1}{R_{a1} + r_{a.t} + (2 + \mu_t)R_c}\right)^{-1}}{1 + g_{m.t}\left(\dfrac{1}{r_{a.t}} + \dfrac{1}{R_{a1} + r_{a.t} + (2 + \mu_t)R_c}\right)^{-1}} \tag{8.20}$$

$$G_{\mu.F.u} = G1_u \, G2_u \tag{8.21}$$

8.4.2 Specific resistances and capacitances C_{c1} & C_{g2}:

8.4.2.1   bypassed version - V1 anode output resistances $R_{o.a.v1.b}$:

$$R_{o.a.v1.b} = r_{a.t} \tag{8.22}$$

8.4.2.2   bypassed version - V2 cathode output resistance $R_{o.c.v2.b}$:

V2 cathode resistance $r_{c.v2}$:

$$r_{c.v2} = \frac{r_{a.t}}{\mu_t + 1} \tag{8.23}$$

V2 cathode load resistance $R1_b$:

$$R1_b = r_{a.t} + R_{c2} + R_{a1} \tag{8.24}$$

output resistance $R_{o.c.v2.b}$:

$$R_{o.c.v2.b} = R1_b \parallel r_{c.v2}$$
$$= \left(\frac{1}{R1_b} + \frac{1}{r_{c.v2}}\right)^{-1} \tag{8.25}$$

8.4.2.3   V1 bypass capacitance C_{c1}:

since R_{v2} is:

$$R_{v2} = r_{a.t} + (\mu_t + 1)(R_{c2} + R_{a1} \parallel R_{g2}) \tag{8.26}$$

V1 cathode resistance $r_{c.v1}$ becomes:

$$r_{c.v1} = \frac{r_{a.t} + R_{v2}}{1 + \mu_t} \tag{8.27}$$

V1 cathode resistance $R_{c.v1}$ becomes:

$$R_{c.v1} = r_{c.v1} \parallel R_c$$
$$= \left(\frac{1}{r_{c.v1}} + \frac{1}{R_c} \right)^{-1} \tag{8.28}$$

and capacitance C_{c1} becomes:

$$C_{c1} = \frac{1}{2\pi f_{c.opt} R_{c.v1}} \tag{8.29}$$

8.4.2.4   V2 input capacitance C_{g2}:

according to Chapter 3.5 the grid's input resistance $R_{in.v2}$ of the V2 gain stage becomes:

$$R_{in.v2} = \frac{R_{g2}}{1 - G2\left(\dfrac{Z1_u - R_c}{Z1_u} \right)} \tag{8.30}$$

$$R1_u = r_{a.t} + (2 + \mu_t) R_c + R_{a1} \tag{8.31}$$

hence, C_{g2} becomes:

$$C_{g2} = \frac{1}{2\pi f_{c.opt} R_{in.v2}} \tag{8.32}$$

8.4.2.5   un-bypassed version - V1 anode output resistance $R_{o.a.v1.u}$:

$$R_{o.a.v1.u} = r_{a.t} + (1 + \mu_t) R_c \tag{8.33}$$

8.4.2.6   un-bypassed version - V2 cathode output resistance $R_{o.c.v2.u}$:

V2 cathode resistance $r_{c.v2}$:

$$r_{c.v2} = \frac{r_{a.t}}{\mu_t + 1} \tag{8.34}$$

output resistance $R_{o.c.v2.b}$:

$$R_{o.c.v2.u} = R1_u \parallel r_{c.v2}$$
$$= \left(\frac{1}{R1_u} + \frac{1}{r_{c.v2}} \right)^{-1} \tag{8.35}$$

8.5 Input capacitances $C_{i.v1}$ and $C_{i.v2}$
Output capacitances $C_{o.v1}$ and $C_{o.v2}$

data sheet figures:
$C_{g.c.v1}$ = grid-cathode capacitance of V1
$C_{g.a.v1}$ = grid-plate capacitance of V1
$C_{a.c.v1}$ = plate-cathode capacitance of V1
$C_{g.c.v2}$ = grid-cathode capacitance of V2
$C_{g.a.v2}$ = grid-plate capacitance of V2
$C_{a.c.v2}$ = plate-cathode capacitance of V2

to be guessed:
$C_{stray.1}$ = sum of stray capacitances around V1
$C_{stray.2}$ = sum of stray capacitances around V2

to be ignored[1]:
$C_{o.c.v2}$ = cathode capacitance of V2

Note: the Miller capacitance of the input depends on V1's gain G1 only[2]!

$$C_{i.v1.b} = \left(1 + |G1_b|\right) C_{g.a.v1} \parallel C_{g.c.v1} \parallel C_{stray.1} \qquad (8.36)$$

$$C_{i.v1.u} = \left(1 + |G1_u|\right) C_{g.a.v1} \parallel C_{g.c.v1} \parallel C_{stray.1} \qquad (8.37)$$

$$C_{i.v2.b} = \left(1 - G2_b\right) C_{g.c.v2} \parallel C_{g.a.v2} \parallel C_{stray.2} \qquad (8.38)$$

$$C_{i.v2.u} = \left(1 - G2_u\right) C_{g.c.v2} \parallel C_{g.a.v2} \parallel C_{stray.2} \qquad (8.39)$$

$$C_{o.v1} = C_{a.c.v1} \parallel C_{c.g.v1} \parallel C_{g.a.v2} \qquad (8.40)$$

$$C_{o.v2} = C_{a.c.v2} \qquad (8.41)$$

8.6 Input impedance Z_{in}

$$Z_{in}(f) = R_{g1} \parallel C_{i.v1} \qquad (8.42)$$

Inclusion of R_{gg} and C_{in} into the calculation course leads to:

$$Z_{in.b}(f) = \left[\left(R_{gg} + \frac{1}{2j\pi f\, C_{i.v1.b}}\right)^{-1} + \frac{1}{R_{g1}}\right]^{-1} + \frac{1}{2j\pi f\, C_{in}} \qquad (8.43)$$

$$Z_{in.u}(f) = \left[\left(R_{gg} + \frac{1}{2j\pi f\, C_{i.v1.u}}\right)^{-1} + \frac{1}{R_{g1}}\right]^{-1} + \frac{1}{2j\pi f\, C_{in}} \qquad (8.44)$$

---

[1]   to be ignored in B_{20k} only because - compared with the size of the corresponding capacitances - the output impedance is rather low

[2]   Miller capacitance effects on C_i and C_o: see Chapter 1.3.2.5

8.7 Gain stage frequency and phase response calculations[3]

8.7.1 Bypassed version:

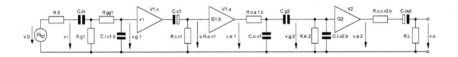

Figure 8.4 Simplified[4] equivalent circuit of Figure 8.1 (bypassed version)
- including all frequency and phase response relevant components

8.7.1.1 1st stage $(V1 = V1_c+V1_a)$ input transfer function $T_{i.1.b}(f)$ and phase $\varphi_{i.1.b}(f)$ -
including source resistance R_S and an oscillation preventing resistor $R_{gg}<<R_g$:

$$T_{i.1.b}(f)=\frac{v_{g.1}}{v_0} \tag{8.45}$$

$$\varphi_{i.1.b}(f)=\arctan\left\{\frac{Im\left[T_{i.1.b}(f)\right]}{Re\left[T_{i.1.b}(f)\right]}\right\} \tag{8.46}$$

$$T_{i.1.b}(f)=\frac{Z2_b(f)\left(\frac{1}{R_{g1}}+\frac{1}{R_{gg1}+Z2_b(f)}\right)^{-1}}{\left(R_{gg1}+Z2_b(f)\right)\left[R_S+Z1(f)+\left(\frac{1}{R_{g1}}+\frac{1}{R_{gg1}+Z2_b(f)}\right)^{-1}\right]} \tag{8.47}$$

$$Z1(f)=\left(2j\pi f\,C_{in}\right)^{-1}$$
$$Z2_b(f)=\left(2j\pi f\,C_{i.v1.b}\right)^{-1} \tag{8.48}$$

8.7.1.2 1st stage $V1_c$ cathode transfer function $T_{c.1.b}(f)$ and phase $\varphi_{c.1.b}(f)$:

$$\varphi_{c.1.b}(f)=\arctan\left\{\frac{Im\left[T_{c.1.b}(f)\right]}{Re\left[T_{c.1.b}(f)\right]}\right\} \tag{8.49}$$

$$T_{c.1.b}(f)=\frac{R_{c.v1}}{R_{c.v1}+\frac{1}{2j\pi f\,C_{c.1}}} \tag{8.50}$$

---

[3]   I've not included into the calculation course a second oscillation preventing resistor $R_{gg.2}$
between $R_{g.2}$ and grid of V2, thus, forming a lp by $R_{gg.2}$ and $C_{i.v2}$. Automatically, this would
mean a further worsening of the phase response at the upper end of B_{20k}.

[4]   See also footnotes 2 and 3 of Chapter 1 and footnote 3 of Chapter 2

8.7.1.3 1st stage output transfer function $T_{o.1.b}(f)$ and phase $\varphi_{o.1.b}(f)$:

$$T_{o.1.b}(f) = \frac{v_{g.2}}{v_{a.1}} \tag{8.51}$$

$$\varphi_{o.1.b}(f) = \arctan\left\{\frac{Im\left[T_{o.1.b}(f)\right]}{Re\left[T_{o.1.b}(f)\right]}\right\} \tag{8.52}$$

$$T_{o.1.b}(f) = \frac{\left[\dfrac{1}{Z3(f)} + \dfrac{1}{Z4(f) + \left(\dfrac{1}{R_{in.2}} + \dfrac{1}{Z5_b(f)}\right)^{-1}}\right]^{-1}}{R_{o.a.v1.b} + \left[\dfrac{1}{Z3(f)} + \dfrac{1}{Z4(f) + \left(\dfrac{1}{R_{in.2}} + \dfrac{1}{Z5_b(f)}\right)^{-1}}\right]^{-1}}$$

$$* \frac{\left(\dfrac{1}{R_{in.2}} + \dfrac{1}{Z5_b(f)}\right)^{-1}}{Z4(f) + \left(\dfrac{1}{R_{in.2}} + \dfrac{1}{Z5_b(f)}\right)^{-1}} \tag{8.53}$$

$$Z3(f) = \left(2j\pi f\, C_{o.v1}\right)^{-1}$$
$$Z4(f) = \left(2j\pi f\, C_{g.2}\right)^{-1} \tag{8.54}$$
$$Z5(f) = \left(2j\pi f\, C_{i.v2.b}\right)^{-1}$$

8.7.1.4 2nd stage output transfer function $T_{o.2.b}(f)$ and phase $\varphi_{o.2.b}(f)$:

$$T_{o.2.b}(f) = \frac{v_o}{v_{c.2}} \tag{8.55}$$

$$\varphi_{o.2.b}(f) = \arctan\left\{\frac{Im\left[T_{o.2.b}(f)\right]}{Re\left[T_{o.2.b}(f)\right]}\right\} \tag{8.56}$$

$$T_{o.2.b}(f) = \frac{R_L}{R_{o.c.v2.b} + R_L + Z6(f)} \tag{8.57}$$

$$Z6(f) = (2j\pi f C_{out})^{-1} \qquad (8.58)$$

8.7.1.5   1st stage fundamental phase shift $\varphi_{G.v1}(f)$:

$$\varphi_{G.v1}(f) = \varphi_{G.\mu.F.b}(f) = -180° \qquad (8.59)$$

8.7.1.6   total gain stage transfer function $T_{tot.b}(f)$ and phase $\varphi_{tot.b}(f)$:

$$T_{tot.b}(f) = T_{i.1.b}(f) T_{c.1.b}(f) T_{o.1.b}(f) T_{o.2.b}(f) G1_b G2_b \qquad (8.60)$$

$$\varphi_{tot.b}(f) = \varphi_{i.1.b}(f) + \varphi_{c.1.b}(f) + \varphi_{o.1.b}(f) + \varphi_{o.2.b}(f) + \varphi_{G.\mu.F.b}(f) \qquad (8.61)$$

8.7.2 Un-bypassed version:

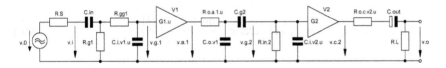

Figure 8.5 Simplified[5] equivalent circuit of Figure 8.1 (un-bypassed version)
- including all frequency and phase response relevant components

8.7.2.1   1st stage (V1) input transfer function $T_{i.1.u}(f)$ and phase $\varphi_{i.1.u}(f)$ -
including source resistance R_S and an oscillation preventing resistor
$R_{gg} \ll R_g$:

$$T_{i.1.u}(f) = \frac{v_{g.1}}{v_0} \qquad (8.62)$$

---

5   See also footnotes 2 and 3 of Chapter 1

$$\varphi_{i.1.u}(f) = \arctan\left\{\frac{\mathrm{Im}\left[T_{i.1.u}(f)\right]}{\mathrm{Re}\left[T_{i.1.u}(f)\right]}\right\} \tag{8.63}$$

$$T_{i.1.u}(f) = \frac{Z2_u(f)\left(\dfrac{1}{R_{g1}} + \dfrac{1}{R_{gg1} + Z2_u(f)}\right)^{-1}}{\left(R_{gg1} + Z2_u(f)\right)\left[R_S + Z1(f) + \left(\dfrac{1}{R_{g1}} + \dfrac{1}{R_{gg1} + Z2_u(f)}\right)^{-1}\right]} \tag{8.64}$$

$$Z1(f) = \left(2j\pi f\,C_{in}\right)^{-1}$$
$$Z2_u(f) = \left(2j\pi f\,C_{i.v1.u}\right)^{-1} \tag{8.65}$$

8.7.2.2   1st stage output transfer function $T_{o.1.u}(f)$ and phase $\varphi_{o.1.u}(f)$:

$$T_{o.1.u}(f) = \frac{v_{g.2}}{v_{a.1}} \tag{8.66}$$

$$\varphi_{o.1.u}(f) = \arctan\left\{\frac{\mathrm{Im}\left[T_{o.1.u}(f)\right]}{\mathrm{Re}\left[T_{o.1.u}(f)\right]}\right\} \tag{8.67}$$

$$T_{o.1.u}(f) = \frac{\left[\dfrac{1}{Z3(f)} + \dfrac{1}{Z4(f) + \left(\dfrac{1}{R_{in.2}} + \dfrac{1}{Z5_u(f)}\right)^{-1}}\right]^{-1}}{R_{o.a.v1.b} + \left[\dfrac{1}{Z3(f)} + \dfrac{1}{Z4(f) + \left(\dfrac{1}{R_{in.2}} + \dfrac{1}{Z5_u(f)}\right)^{-1}}\right]^{-1}}$$
$$*\ \frac{\left(\dfrac{1}{R_{in.2}} + \dfrac{1}{Z5_u(f)}\right)^{-1}}{Z4(f) + \left(\dfrac{1}{R_{in.2}} + \dfrac{1}{Z5_u(f)}\right)^{-1}} \tag{8.68}$$

$$Z3(f) = (2j\pi f C_{o.v1})^{-1}$$

$$Z4(f) = (2j\pi f C_{g.2})^{-1} \tag{8.69}$$

$$Z5(f) = (2j\pi f C_{i.v2.b})^{-1}$$

8.7.2.3 2nd stage output transfer function $T_{o.2.u}(f)$ and phase $\varphi_{o.2.u}(f)$:

$$T_{o.2.u}(f) = \frac{v_o}{v_{c.2}} \tag{8.70}$$

$$\varphi_{o.2.u}(f) = \arctan\left\{\frac{\mathrm{Im}\left[T_{o.2.u}(f)\right]}{\mathrm{Re}\left[T_{o.2.u}(f)\right]}\right\} \tag{8.71}$$

$$T_{o.2.u}(f) = \frac{R_L}{R_{o.c.v2.u} + R_L + Z6(f)} \tag{8.72}$$

$$Z6(f) = (2j\pi f C_{out})^{-1} \tag{8.73}$$

8.7.2.4 1st stage fundamental phase shift $\varphi_{G.v1}(f)$:

$$\varphi_{G.v1}(f) = \varphi_{G.\mu.F.u}(f) = -180° \tag{8.74}$$

8.7.2.5   total gain stage transfer function $T_{tot.u}(f)$ and phase $\varphi_{tot.u}(f)$:

$$T_{tot.u}(f) = T_{i.1.u}(f) T_{o.1.u}(f) T_{o.2.u}(f) G1_u G2_u \tag{8.75}$$

$$\varphi_{tot.u}(f) = \varphi_{i.1.u}(f) + \varphi_{o.1.u}(f) + \varphi_{o.2.u}(f) + \varphi_{G.\mu.F.u}(f) \tag{8.76}$$

8.8   Example with ECC83 / 12AX7 (83):

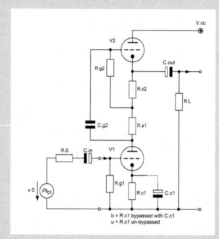

Figure 8.6 µ-Follower example circuitry

8.8.1   Triode bias data:

$$I_{a1.83} = I_{a2.83} = 1.2 \cdot 10^{-3} A \qquad V_{a1.83} = 100V \qquad V_{a2.83} = 100V$$

$$V_{cc} = 268.5V \qquad V_{g1.83} = -0.5V \qquad V_{g2.83} = 168V$$

8.8.2   Triode valve constants:

$$g_{m.83} := 1.75 \cdot 10^{-3} \cdot S \qquad \mu_{83} := 101.5 \qquad r_{a.83} := 58 \cdot 10^{3} \Omega$$

$$g_{m.v1} := g_{m.83} \qquad \mu_{v1} := \mu_{83} \qquad r_{a.v1} := r_{a.83}$$

$$g_{m.v2} := g_{m.83} \qquad \mu_{v2} := \mu_{83} \qquad r_{a.v2} := r_{a.83}$$

$$C_{g.c.1.83} := 1.65 \cdot 10^{-12} F \qquad C_{g.a.1.83} := 1.6 \cdot 10^{-12} F \qquad C_{a.c.1.83} := 0.33 \cdot 10^{-12} F$$

$$C_{g.c.2.83} := 1.65 \cdot 10^{-12} F \qquad C_{g.a.2.83} := 1.6 \cdot 10^{-12} F \qquad C_{a.c.2.83} := 0.33 \cdot 10^{-12} F$$

8.8.3   Circuit variables:

$$R_{c1.83} := 417\Omega \qquad R_{c2.83} := R_{c1.83} \qquad R_{c.83} := R_{c1.83}$$

$$R_S := 1 \cdot 10^{3} \Omega \qquad R_L := 47.5 \cdot 10^{3} \Omega \qquad R_{gg1.83} := 0.22 \cdot 10^{3} \Omega$$

$$R_{g1.83} := 47.5 \cdot 10^{3} \Omega \qquad C_{in.83} := 2.2 \cdot 10^{-6} F \qquad C_{out.83} := 10 \cdot 10^{-6} F$$

$$R_{g2.83} := 1 \cdot 10^{6} \Omega \qquad C_{stray.83.1} := 10 \cdot 10^{-12} F \qquad C_{stray.83.2} := 10 \cdot 10^{-12} F$$

$$R_{a1.83} := 56.2 \cdot 10^{3} \Omega$$

➢   MCD Worksheet VIII   μ-F calculations   page 2

8.8.4 Calculation relevant data:

frequency range f for the below shown graphs: $f := 10\,Hz, 20\,Hz .. 100000\,Hz$

$h := 1000 \cdot Hz$

8.8.5 Gain $G_{\mu.F83.b}$:

$$R_{v2} := r_{a.v2} + \left(1 + \mu_{v2}\right) \cdot \left[R_{c2.83} + \left(\frac{1}{R_{a1.83}} + \frac{1}{R_{g2.83}} \right)^{-1} \right]$$

$R_{v2} = 5.555 \times 10^6\,\Omega$

$$Gl_b := -g_{m.83} \cdot \left(\frac{1}{R_{v2}} + \frac{1}{r_{a.83}} \right)^{-1}$$

via V1 $Gl_b = -100.451 \times 10^0$

$$G2_b := \frac{g_{m.v2} \cdot \left(\frac{1}{r_{a.v2}} + \frac{1}{R_{c2.83} + R_{a1.83} + r_{a.v1}} \right)^{-1}}{1 + g_{m.v2} \cdot \left(\frac{1}{r_{a.v2}} + \frac{1}{R_{c2.83} + R_{a1.83} + r_{a.v1}} \right)^{-1}}$$

via V2 $G2_b = 985.379 \times 10^{-3}$

$G_{\mu.F.83.b} := Gl_b \cdot G2_b$ $G_{\mu.F.83.b} = -98.982 \times 10^0$

$G_{\mu.F.83.b.e} := 20 \cdot \log\left(\left| G_{\mu.F.83.b} \right| \right)$ $G_{\mu.F.83.b.e} = 39.911 \times 10^0$ [dB]

$G_{\mu.F.83.b.rot} := \mu_{83}$ $G_{\mu.F.83.b.rot} = 101.5 \times 10^0$

$G_{\mu.F.83.b.rot.e} := 20 \cdot \log\left(\left| G_{\mu.F.83.b.rot} \right| \right)$ $G_{\mu.F.83.b.rot.e} = 40.129 \times 10^0$ [dB]

$$G_{\mu.F.83.b.e} - G_{\mu.F.83.b.rot.e} = -218.156 \times 10^{-3}$$

8.8.6 Gain $G_{\mu.F83.u}$:

$$Gl_u := -g_{m.83} \cdot \frac{r_{a.83} \cdot R_{v2}}{R_{v2} + r_{a.83} + R_{c.83} \cdot \left(1 + \mu_{83}\right)}$$

$Gl_u = -99.692 \times 10^0$

$$G2_u := \frac{g_{m.v2} \cdot \left[\frac{1}{r_{a.v2}} + \frac{1}{R_{c2.83} + R_{a1.83} + r_{a.v1} + \left(1 + \mu_{v1}\right) \cdot R_{c1.83}} \right]^{-1}}{1 + g_{m.v2} \cdot \left[\frac{1}{r_{a.v2}} + \frac{1}{R_{c2.83} + R_{a1.83} + r_{a.v1} + \left(1 + \mu_{v1}\right) \cdot R_{c1.83}} \right]^{-1}}$$

$G2_u = 986.696 \times 10^{-3}$

> ➤   MCD Worksheet VIII   μ-F calculations   page 3

$G_{\mu.F.83.u} := G1_u \cdot G2_u$ $G_{\mu.F.83.u} = -98.366 \times 10^0$

$G_{\mu.F.83.u.e} := 20 \cdot \log\left(\left|G_{\mu.F.83.u}\right|\right)$ $G_{\mu.F.83.u.e} = 39.857 \times 10^0$ [dB]

$G_{\mu.F.83.u.rot} := \mu_{83}$ $G_{\mu.F.83.u.rot} = 101.5 \times 10^0$

$G_{\mu.F.83.u.rot.e} := 20 \cdot \log\left(\left|G_{\mu.F.83.u.rot}\right|\right)$ $G_{\mu.F.83.u.rot.e} = 40.129 \times 10^0$ [dB]

$$G_{\mu.F.83.u.e} - G_{\mu.F.83.u.rot.e} = -272.453 \times 10^{-3}$$

8.8.7   Specific resistances:

$R_{o.a.v1.b} := r_{a.v1}$ $R_{o.a.v1.b} = 58 \times 10^3\,\Omega$

$R_{o.a.v1.u} := r_{a.v1} + \left(1 + \mu_{v1}\right) \cdot R_{c1.83}$ $R_{o.a.v1.u} = 100.743 \times 10^3\,\Omega$

$r_{c.v1} := \dfrac{r_{a.v1} + R_{v2}}{1 + \mu_{v1}}$ $r_{c.v1} = 54.758 \times 10^3\,\Omega$

$R_{c.v1} := \left(\dfrac{1}{r_{c.v1}} + \dfrac{1}{R_{c1.83}}\right)^{-1}$ $R_{c.v1} = 413.848 \times 10^0\,\Omega$

$r_{c.v2} := \dfrac{r_{a.v2}}{1 + \mu_{v2}}$ $r_{c.v2} = 565.854 \times 10^0\,\Omega$

$R1_b := r_{a.v1} + R_{c2.83} + \left(\dfrac{1}{R_{a1.83}} + \dfrac{1}{R_{g2.83}}\right)^{-1}$ $R1_b = 111.627 \times 10^3\,\Omega$

$R_{o.c.v2.b} := \left(\dfrac{1}{R1_b} + \dfrac{1}{r_{c.v2}}\right)^{-1}$ $R_{o.c.v2.b} = 563 \times 10^0\,\Omega$

$R1_u := r_{a.v1} + \left(1 + \mu_{v1}\right) \cdot R_{c1.83} + R_{c2.83} + \left(\dfrac{1}{R_{a1.83}} + \dfrac{1}{R_{g2.83}}\right)^{-1}$ $R1_u = 154.369 \times 10^3\,\Omega$

$R_{o.c.v2.u} := \left(\dfrac{1}{R1_u} + \dfrac{1}{r_{c.v2}}\right)^{-1}$ $R_{o.c.v2.u} = 563.787 \times 10^0\,\Omega$

8.8.8   Capacitance C_{c1} and C_{g2}:

$f_c := 20\text{Hz}$ $f_{c.opt} := \dfrac{f_c}{100}$ $f_{c.opt} = 200 \times 10^{-3}\,\text{Hz}$

$C_{c1} := \dfrac{1}{2 \cdot \pi \cdot f_{c.opt} \cdot R_{c.v1}}$ $C_{c1} = 1.923 \times 10^{-3}\,\text{F}$

$$R_{in.v2} := \frac{R_{g2.83}}{1 - G2_u \cdot \frac{R1_u - R_{c2.83}}{R1_u}} \qquad\qquad R_{in.v2} = 62.619 \times 10^6\,\Omega$$

$$C_{g2} := \frac{1}{2 \cdot \pi \cdot f_{c.opt} \cdot R_{in.v2}} \qquad\qquad C_{g2} = 12.708 \times 10^{-9}\,F$$

8.8.9 Specific input and output capacitances :

$$C_{i.v1.b} := \left(1 + |G1_b|\right) \cdot C_{g.a.1.83} + C_{g.c.1.83} + C_{stray.83.1} \qquad C_{i.v1.b} = 173.972 \times 10^{-12}\,F$$

$$C_{i.v1.u} := \left(1 + |G1_u|\right) \cdot C_{g.a.1.83} + C_{g.c.1.83} + C_{stray.83.1} \qquad C_{i.v1.u} = 172.757 \times 10^{-12}\,F$$

$$C_{i.v2.b} := \left(1 - G2_b\right) \cdot C_{g.a.2.83} + C_{g.c.2.83} + C_{stray.83.2} \qquad C_{i.v2.b} = 11.673 \times 10^{-12}\,F$$

$$C_{i.v2.u} := \left(1 - G2_u\right) \cdot C_{g.a.2.83} + C_{g.c.2.83} + C_{stray.83.2} \qquad C_{i.v2.u} = 11.671 \times 10^{-12}\,F$$

$$C_{o.v1} := C_{a.c.1.83} + C_{g.a.1.83} + C_{g.a.2.83} \qquad\qquad C_{o.v2} := C_{a.c.2.83}$$

8.8.10 Gain stage frequency and phase response
- bypassed version:

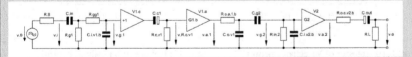

Figure 8.7 = Figure 8.4

$$Z1(f) := \frac{1}{2j \cdot \pi \cdot f \cdot C_{in.83}} \qquad\qquad Z2_b(f) := \frac{1}{2j \cdot \pi \cdot f \cdot C_{i.v1.b}}$$

$$T_{i.1.b}(f) := \frac{Z2_b(f) \cdot \left(\frac{1}{R_{g1.83}} + \frac{1}{R_{gg1.83} + Z2_b(f)}\right)^{-1}}{\left(Z2_b(f) + R_{gg1.83}\right) \cdot \left[R_S + Z1(f) + \left(\frac{1}{R_{g1.83}} + \frac{1}{R_{gg1.83} + Z2_b(f)}\right)^{-1}\right]}$$

$$T_{i.1.b.e}(f) := 20 \cdot \log\left(\left|T_{i.1.b}(f)\right|\right) \qquad\qquad \phi_{i.1.b}(f) := atan\left(\frac{Im\left(T_{i.1.b}(f)\right)}{Re\left(T_{i.1.b}(f)\right)}\right)$$

➢   MCD Worksheet VIII   μ-F calculations   page 5

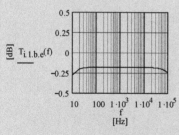

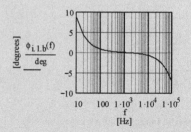

Figure 8.8 Transfer of i/p network   Figure 8.9 Phase of i/p network

$$T_{c.1.b}(f) := \left(\frac{R_{c.v1}}{R_{c.v1} + \dfrac{1}{2j \cdot \pi \cdot f \cdot C_{c1}}} \right)$$

$$\phi_{c.1.b}(f) := atan\left(\frac{Im\left(T_{c.1.b}(f)\right)}{Re\left(T_{c.1.b}(f)\right)} \right)$$

$$T_{c.1.b.e}(f) := 20 \cdot log\left(\left| T_{c.1.b}(f) \right| \right)$$

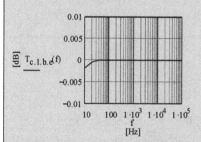

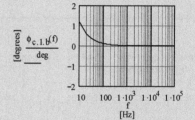

Figure 8.10 Transfer of V1 cathode network   Figure 8.11 Phase of V1 cathode network

$$Z3(f) := \frac{1}{2j \cdot \pi \cdot f \cdot C_{o.v1}} \qquad Z4(f) := \frac{1}{2j \cdot \pi \cdot f \cdot C_{g2}} \qquad Z5_b(f) := \frac{1}{2j \cdot \pi \cdot f \cdot C_{i.v2.b}}$$

$$T_{o.1.b}(f) := \frac{\left[\dfrac{1}{Z3(f)} + \dfrac{1}{Z4(f) + \left(\dfrac{1}{R_{in.v2}} + \dfrac{1}{Z5_b(f)} \right)^{-1}} \right]^{-1}}{R_{o.a.v1.b} + \left[\dfrac{1}{Z3(f)} + \dfrac{1}{Z4(f) + \left(\dfrac{1}{R_{in.v2}} + \dfrac{1}{Z5_b(f)} \right)^{-1}} \right]^{-1}} \cdot \frac{\left(\dfrac{1}{R_{in.v2}} + \dfrac{1}{Z5_b(f)} \right)^{-1}}{Z4(f) + \left(\dfrac{1}{R_{in.v2}} + \dfrac{1}{Z5_b(f)} \right)^{-1}}$$

$$\phi_{o.1.b}(f) := atan\left(\frac{Im\left(T_{o.1.b}(f)\right)}{Re\left(T_{o.1.b}(f)\right)} \right)$$

$$T_{o.1.b.e}(f) := 20 \cdot log\left(\left| T_{o.1.b}(f) \right| \right)$$

➤   MCD Worksheet VIII   μ-F calculations   page 6

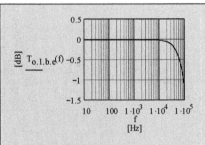

Figure 8.12 Transfer of V1 o/p network   Figure 8.13 Phase of V1 o/p network

$$Z6(f) := \frac{1}{2j \cdot \pi \cdot f \cdot C_{out.83}}$$

$$T_{o.2.b.e}(f) := 20 \cdot \log\left(\left|T_{o.2.b}(f)\right|\right)$$

$$T_{o.2.b}(f) := \frac{R_L}{R_{o.c.v2.b} + R_L + Z6(f)}$$

$$\phi_{o.2.b}(f) := atan\left(\frac{Im\left(T_{o.2.b}(f)\right)}{Re\left(T_{o.2.b}(f)\right)}\right)$$

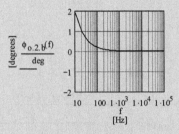

Figure 8.14 Transfer of V2 o/p network   Figure 8.15 Phase of V2 o/p network

$$T_{tot.83.b}(f) := T_{i.1.b}(f) \cdot T_{c.1.b}(f) \cdot T_{o.1.b}(f) \cdot T_{o.2.b}(f) \cdot G1_b \cdot G2_b$$

$$T_{tot.83.b.e}(f) := 20 \cdot \log\left(\left|T_{tot.83.b}(f)\right|\right) \qquad\qquad \phi_{G.83}(f) := -180 \, deg$$

$$\phi_{tot.83.b}(f) := \phi_{i.1.b}(f) + \phi_{c.1.b}(f) + \phi_{o.1.b}(f) + \phi_{o.2.b}(f) + \phi_{G.83}(f)$$

> MCD Worksheet VIII   μ-F calculations   page 7

8.8.11 Gain stage frequency and phase response
- un-bypassed version:

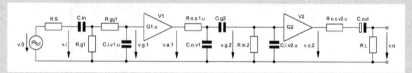

Figure 8.16 = Figure 8.5

$$Z2_u(f) := \frac{1}{2j \cdot \pi \cdot f \cdot C_{i.v1.u}}$$

$$T_{i.1.u}(f) := \frac{Z2_u(f) \cdot \left(\dfrac{1}{R_{g1.83}} + \dfrac{1}{R_{gg1.83} + Z2_u(f)} \right)^{-1}}{\left(Z2_u(f) + R_{gg1.83} \right) \cdot \left[R_S + Z1(f) + \left(\dfrac{1}{R_{g1.83}} + \dfrac{1}{R_{gg1.83} + Z2_u(f)} \right)^{-1} \right]}$$

$$Z5_u(f) := \frac{1}{2j \cdot \pi \cdot f \cdot C_{i.v2.u}} \qquad\qquad \phi_{i.1.u}(f) := atan\left(\frac{Im\left(T_{i.1.u}(f) \right)}{Re\left(T_{i.1.u}(f) \right)} \right)$$

$$T_{o.1.u}(f) := \frac{\left[\dfrac{1}{Z3(f)} + \dfrac{1}{Z4(f) + \left(\dfrac{1}{R_{in.v2}} + \dfrac{1}{Z5_u(f)} \right)^{-1}} \right]^{-1}}{R_{o.a.v1.u} + \left[\dfrac{1}{Z3(f)} + \dfrac{1}{Z4(f) + \left(\dfrac{1}{R_{in.v2}} + \dfrac{1}{Z5_u(f)} \right)^{-1}} \right]^{-1}} \cdot \frac{\left(\dfrac{1}{R_{in.v2}} + \dfrac{1}{Z5_u(f)} \right)^{-1}}{Z4(f) + \left(\dfrac{1}{R_{in.v2}} + \dfrac{1}{Z5_u(f)} \right)^{-1}}$$

$$\phi_{o.1.u}(f) := atan\left(\frac{Im\left(T_{o.1.u}(f) \right)}{Re\left(T_{o.1.u}(f) \right)} \right)$$

$$T_{o.2.u}(f) := \frac{R_L}{R_{o.c.v2.u} + R_L + Z6(f)} \qquad\qquad \phi_{o.2.u}(f) := atan\left(\frac{Im\left(T_{o.2.u}(f) \right)}{Re\left(T_{o.2.u}(f) \right)} \right)$$

$$T_{tot.83.u}(f) := T_{i.1.u}(f) \cdot T_{o.1.u}(f) \cdot T_{o.2.u}(f) \cdot G1_u \cdot G2_u$$

$$T_{tot.83.u.e}(f) := 20 \cdot log\left(\left| T_{tot.83.u}(f) \right| \right) \qquad\qquad \phi_{G.83}(f) := -180deg$$

$$\phi_{tot.83.u}(f) := \phi_{i.1.u}(f) + \phi_{o.1.u}(f) + \phi_{o.2.u}(f) + \phi_{G.83}(f)$$

> ➤ MCD Worksheet VIII   μ-F calculations   page 8

8.8.12 Frequency and phase response plots:

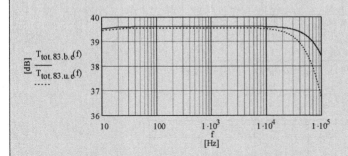

Figure 8.17 Frequency responses of the bypassed and un-bypassed
μ-F gain stage

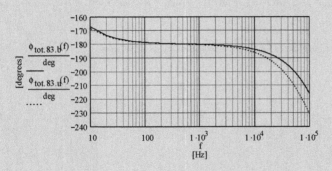

Figure 8.18 Phase responses of the bypassed and un-bypassed
μ-F gain stage

8.8.13 Input impedances:

$$Z_{in.b}(f) := \left[\left(R_{gg1.83} + \frac{1}{2j \cdot \pi \cdot f \cdot C_{i.vl.b}} \right)^{-1} + \frac{1}{R_{g1.83}} \right]^{-1} + \frac{1}{2j \cdot \pi \cdot f \cdot C_{in.83}}$$

$$Z_{in.u}(f) := \left[\left(R_{gg1.83} + \frac{1}{2j \cdot \pi \cdot f \cdot C_{i.vl.u}} \right)^{-1} + \frac{1}{R_{g1.83}} \right]^{-1} + \frac{1}{2j \cdot \pi \cdot f \cdot C_{in.83}}$$

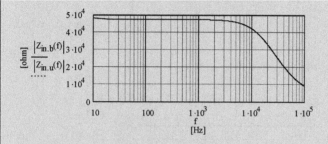

Figure 8.19 Input impedances for the bypassed and un-bypassed versions

Chapter 9 The Aikido Gain Stage (AIK)

9.1 Circuit diagram

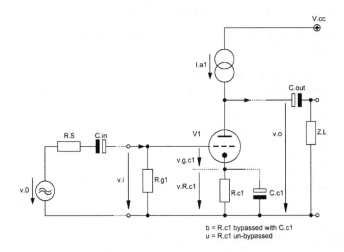

Figure 9.1 Basic design of the AIK Gain Stage

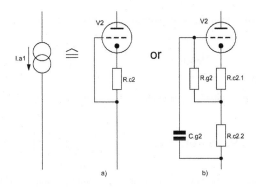

Figure 9.2 Current source alternatives: a) = low-Z (lo) and b) = high-Z (hi)[1]

---

[1] Inclusion of an oscillation preventing resistor R_{gg2} [between R_{g2} and grid of V2 alternative b)] into the calculation course, thus, forming a lp by R_{gg2} and $C_{i.v2}$, will automatically lead to a worsening of the frequency and phase response at the upper end of B_{20k}.

9.2   Basic assumptions

Generally, the whole AIK gain stage consists of a CCS or CCS+Cc gain stage (V1 with gain G) of Chapters 1 or 2 with an active high impedance (high-Z) or low impedance (low-Z) current source (V2) of Chapters 5.1 or 5.3 as the plate load. The output signal of the gain stage is taken from the plate of V1, thus, the output impedance is rather high and will react highly sensitive on rather low load impedances R_L^2 or Z_L.

To get all relevant formulae shown further down these lines the following assumption were made:

- V1 equals V2 (ideal case: double triode)
- plate current V1   = plate current V2
- $g_{m.v1}$ $= g_{m.v2}$ $= g_{m.t}$
- $r_{a.v1}$ $= r_{a.v2}$ $= r_{a.t}$
- μ_{v1} $= \mu_{v2}$ $= \mu_t$
- $R_{c2.1}$ $= R_{c1}$ $= R_c$
- $R_{c2.2} > 0R$
- bypassed version (b): R_{c1} = bypassed by C_{c1}
- un-bypassed version (u): R_{c1} = un-bypassed

9.3   Gains $G_{aik.lo}$ for V1 with the low-Z plate load[3]

The upper valve's active AC resistance $R_{v2.lo}$ becomes:

$$R_{v2.lo} = r_{a.t} + (1 + \mu_t) R_{c2} \tag{9.1}$$

9.3.1 bypassed version - gain $G_{aik.b.lo}$ in terms of $g_{m.t}$ - excl. R_L-effect:

$$G_{aik.b.lo} = -g_{m.t} \left(\frac{1}{r_{a.t}} + \frac{1}{R_{v2.lo}} \right) \tag{9.2}$$

9.3.2 bypassed version - gain $G_{aik.b.lo.eff}$ in terms of $g_{m.t}$ - incl. R_L-effect:

$$G_{aik.b.lo.eff} = -g_{m.t} \left(\frac{1}{r_{a.t}} + \frac{1}{R_L} + \frac{1}{R_{v2.lo}} \right)^{-1} \tag{9.3}$$

9.3.3 un-bypassed version - gain $G_{aik.u.lo}$ in terms of $g_{m.t}$ and μ_t - excl. R_L-effect:

$$G_{aik.u.lo} = -\frac{1}{2} g_{m.t} r_{a.t}$$
$$= -\frac{1}{2} \mu_t \tag{9.4}$$

---

[2]   see respective plots (effective gains vs. R_L changes) on MCD worksheets IX-1 and IX-2
[3] R_L is the resistive part of Z_L in Figure 9.1

9.3.4 un-bypassed version - gain $G_{aik.u.lo.eff}$ in terms of $g_{m.t}$ and μ_t - incl. R_L-effect:

$$G_{aik.u.lo.eff} = -g_{m.t} \frac{r_{a.t} R_L}{r_{a.t} + (1 + \mu_t) R_c + 2R_L}$$

$$= -\mu_t \frac{R_L}{r_{a.t} + (1 + \mu_t) R_c + 2R_L} \tag{9.5}$$

9.4   Gains $G_{aik.hi}$ for V1 with the high-Z plate load

The upper valve's active AC resistance $R_{v2.hi}$ becomes:

$$R_{v2.hi} = r_{a.t} + (1 + \mu_t) \left[R_{c2.1} + \left(R_{c2.2} \parallel R_{g2} \right) \right] \tag{9.6}$$

9.4.1 bypassed version - gain $G_{aik.b.hi}$ in terms of $g_{m.t}$ - excl. R_L-effect:

$$G_{aik.b.hi} = -g_{m.t} \left(\frac{1}{r_{a.t}} + \frac{1}{R_{v2.hi}} \right) \tag{9.7}$$

9.4.2 bypassed version - gain $G_{aik.b.hi.eff}$ in terms of $g_{m.t}$ - incl. R_L-effect:

$$G_{aik.b.hi.eff} = -g_{m.t} \left(\frac{1}{r_{a.t}} + \frac{1}{R_L} + \frac{1}{R_{v2.hi}} \right)^{-1} \tag{9.8}$$

9.4.3 un-bypassed version - gain $G_{aik.u.hi}$ in terms of $g_{m.t}$ and μ_t - excl. R_L-effect:

$$G_{aik.u.hi} = -g_{m.t} \frac{r_{a.t} R_{v2.hi}}{r_{a.t} + (1 + \mu_t) R_c + R_{v2.hi}}$$

$$= -\mu_t \frac{R_{v2.hi}}{r_{a.t} + (1 + \mu_t) R_c + R_{v2.hi}} \tag{9.9}$$

9.4.4 un-bypassed version - gain $G_{aik.u.hi.eff}$ in terms of $g_{m.t}$ and μ_t - incl. R_L-effect:

$$G_{aik.u.hi.eff} = -g_{m.t} \frac{r_{a.t} \left(\dfrac{1}{R_{v2.hi}} + \dfrac{1}{R_L} \right)^{-1}}{r_{a.t} + (1 + \mu_t) R_c + \left(\dfrac{1}{R_{v2.hi}} + \dfrac{1}{R_L} \right)^{-1}}$$

$$= -\mu_t \frac{\left(\dfrac{1}{R_{v2.hi}} + \dfrac{1}{R_L} \right)^{-1}}{r_{a.t} + (1 + \mu_t) R_c + \left(\dfrac{1}{R_{v2.hi}} + \dfrac{1}{R_L} \right)^{-1}} \tag{9.10}$$

9.5   Derivations

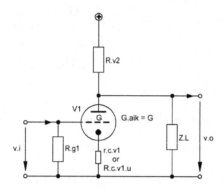

Figure 9.3 Simplified equivalent circuit of Figure 9.1

9.5.1 Gains[4]:

Because all relevant calculation formulae were given in the previous chapters we can take the much easier to handle Figure 9.3 as the basis of the derivations. It shows that the AIK gain stage is made of one gain block around V1 with gain G only. G_b is a CCS+Cc gain stage with AC-grounded cathode à la Chapter 2, G_u is a CCS gain stage à la Chapter 1 and the active plate load is a low-Z (R_{c2}!) or high-Z ($R_{c2.1}$ plus $R_{c2.2}$!) current source à la Chapters 5.1 and 5.3. Without big derivation efforts the respective formulae can be used to calculate all relevant gains $G_{aik.b}$ or $G_{aik.u}$ - also by inclusion of all R_L-effects (= $_{eff}$) and Z_L-effects into the whole calculation course.

9.5.1.1   bypassed lo version:

$G_{aik.b.lo}$ and $G_{aik.b.lo.eff}$ become[5]:

$$G1_{aik.b.lo} = - g_{m.t} \left(R_{v2.lo} \parallel r_{a.t} \right) \tag{9.11}$$

$$G1_{aik.b.lo.eff} = - g_{m.t} \left(R_{v2.lo} \parallel r_{a.t} \parallel R_L \right) \tag{9.12}$$

$$\begin{aligned} R_{v2.lo} &= r_{a.t} + \left(1 + g_{m.t} r_{a.t}\right) R_c \\ &= r_{a.t} + \left(1 + \mu_t\right) R_c \end{aligned} \tag{9.13}$$

$$\begin{aligned} G_{aik.b.lo} &= - g_{m.t} \frac{r_{a.t} \left[r_{a.t} + \left(1 + g_{m.t} r_{a.t}\right) R_c \right]}{2 r_{a.t} + \left(1 + g_{m.t} r_{a.t}\right) R_c} \\ &= - \mu_t \frac{r_{a.t} + \left(1 + \mu_t\right) R_c}{2 r_{a.t} + \left(1 + \mu_t\right) R_c} \end{aligned} \tag{9.14}$$

---

[4]   no rules of thumb are given in this section because of the high impact of the o/p load Z_L on the gains (see MCD IX worksheets)

[5]   see derivation of respective formula of Chapter 2

$$G_{aik.b.lo.eff} = -g_{m.t}\left(\frac{1}{r_{a.t}} + \frac{1}{R_{v2.lo}} + \frac{1}{R_L}\right)^{-1} \tag{9.15}$$

9.5.1.2   un-bypassed lo version:

$G_{aik.u.lo}$ and $G_{aik.u.lo.eff}$ become[6]:

$$G_{aik.u.lo} = -g_{m.t}\frac{r_{a.t}\,R_{v2.lo}}{r_{a.t} + R_{v2.lo} + (1 + g_{m.t}\,r_{a.t})R_c} \tag{9.16}$$

$$\Rightarrow G_{aik.u.lo} = -\frac{1}{2}g_{m.t}\,r_{a.t} \tag{9.17}$$
$$= -\frac{1}{2}\mu_t$$

$$G_{aik.u.lo.eff} = -g_{m.t}\frac{r_{a.t}\left(\dfrac{1}{R_{v2.lo}} + \dfrac{1}{R_L}\right)^{-1}}{r_{a.t} + (1 + g_{m.t}\,r_{a.t})R_c + \left(\dfrac{1}{R_{v2.lo}} + \dfrac{1}{R_L}\right)^{-1}} \tag{9.18}$$

$$\Rightarrow G_{aik.b.lo.eff} = -g_{m.t}\frac{r_{a.t}R_L}{r_{a.t} + (1+\mu_t)R_c + 2R_L} \tag{9.19}$$
$$= \mu_t\frac{R_L}{r_{a.t} + (1+\mu_t)R_c + 2R_L}$$

9.5.1.3   bypassed hi version[7]:

$G_{aik.b.hi}$ and $G_{aik.b.hi.eff}$ become:

$$G1_{aik.b.hi} = -g_{m.t}\left(R_{v2.hi} \parallel r_{a.t}\right) \tag{9.20}$$

$$G1_{aik.b.hi.eff} = -g_{m.t}\left(R_{v2.hi} \parallel r_{a.t} \parallel R_L\right) \tag{9.21}$$

$$R_{v2.hi} = r_{a.t} + (1+\mu_t)\left[R_{c2.1} + \left(\frac{1}{R_{c2.2}} + \frac{1}{R_{g2}}\right)^{-1}\right] \tag{9.22}$$

$$\Rightarrow G_{aik.b.hi} = -g_{m.t}\left(\frac{1}{r_{a.t}} + \frac{1}{R_{v2.hi}}\right)^{-1} \tag{9.23}$$

---

[6]   see derivation of respective formula of Chapter 1
[7]   see derivation of respective formula of Chapter 2

$$\Rightarrow G_{aik.b.hi.eff} = -g_{m.t} \left(\frac{1}{r_{a.t}} + \frac{1}{R_{v2.hi}} + \frac{1}{R_L} \right)^{-1} \qquad (9.24)$$

9.5.1.4 un-bypassed hi version:

$G_{aik.u.hi}$ and $G_{aik.u.hi.eff}$ become[8]:

$$\Rightarrow G_{aik.u.hi} = -g_{m.t} \frac{r_{a.t} R_{v2.hi}}{r_{a.t} + (1 + g_{m.t} r_{a.t})R_c + R_{v2.hi}}$$

$$= \mu_t \frac{R_{v2.hi}}{r_{a.t} + (1 + \mu_t)R_c + R_{v2.hi}} \qquad (9.25)$$

$$\Rightarrow G_{aik.u.hi.eff} = -g_{m.t} \frac{r_{a.t} \left(\frac{1}{R_{v2.hi}} + \frac{1}{R_L} \right)^{-1}}{r_{a.t} + (1 + g_{m.t} r_{a.t})R_c + \left(\frac{1}{R_{v2.hi}} + \frac{1}{R_L} \right)^{-1}}$$

$$= -\mu_t \frac{\left(\frac{1}{R_{v2.hi}} + \frac{1}{R_L} \right)^{-1}}{r_{a.t} + (1 + \mu_t)R_c + \left(\frac{1}{R_{v2.hi}} + \frac{1}{R_L} \right)^{-1}} \qquad (9.26)$$

9.5.2 Specific resistances, impedances and capacitance C_{c1}:

9.5.2.1 the active load resistances $R_{v2.lo}$ and $R_{v2.hi}$ become[9]:

$$R_{v2.lo} = r_{a.t} + (\mu_t + 1) R_{c2} \qquad (9.27)$$

$$R_{v2.hi} = r_{a.t} + (1 + \mu_t) \left[R_{c2.1} + \left(\frac{1}{R_{c2.2}} + \frac{1}{R_{g2}} \right)^{-1} \right] \qquad (9.28)$$

the active load impedances $Z_{v2.lo}(f)$ and $Z_{v2.hi}(f)$ become:

$$Z_{v2.lo}(f) = \left[(r_{a.t} \parallel C_{a.c.v2}) + (1 + \mu_t)(R_{c2} \parallel C_{g.c.v2}) \right] \parallel C_{g.a.v2} \parallel C_{stray.2} \qquad (9.29)$$

---

[8]   see derivation of respective formula of Chapter 1
[9]   see Chapter 5.1 and 5.3

$$Z_{v2.hi}(f) = \left(\begin{cases} \begin{cases} \left(r_{a.t} \parallel C_{a.c.v2}\right) + \\ \left(1 + \mu_t\right)\left[\left\{R_{c2.1} + \left(R_{c2.2} \parallel R_{g2}\right)\right\} \parallel C_{g.c.v2}\right] \end{cases} \\ \parallel C_{g.a.v2} \parallel C_{stray.2} \end{cases} \right) \tag{9.30}$$

9.5.2.2   bypassed version - V1 plate output resistance $R_{o.a.v1.b}$
and impedance $Z_{o.a.v1.b}(f)$:

$$R_{o.a.v1.b} = r_{a.t} \tag{9.31}$$

$$Z_{o.a.v1.b}(f) = R_{o.a.v1.b} \parallel C_{o.v1} \tag{9.32}$$

9.5.2.3   bypassed version - AIK gain stage o/p resistances $R_{o.aik.b.lo}$ and $R_{o.aik.b.hi}$ and o/p
impedances $Z_{o.aik.b.lo}(f)$ and $Z_{o.aik.b.hi}(f)$ at the plate of V1:

$$R_{o.aik.b.lo} = R_{o.a.v1.b} \parallel R_{v2.lo} \tag{9.33}$$

$$R_{o.aik.b.hi} = R_{o.a.v1.b} \parallel R_{v2.hi} \tag{9.34}$$

$$Z_{o.aik.b.lo}(f) = Z_{o.a.v1.b}(f) \parallel Z_{v2.lo}(f) \tag{9.35}$$

$$Z_{o.aik.b.hi}(f) = Z_{o.a.v1.b}(f) \parallel Z_{v2.hi}(f) \tag{9.36}$$

9.5.2.4   V1 cathode resistor bypass capacitances $C_{c1.lo}$ and $C_{c1.hi}$ become:

with V1 cathode output resistances $r_{c.v1.lo}$ and $r_{c.v1.hi}$:

$$r_{c.v1.lo} = \frac{r_{a.t} + Z_{v2.lo}}{1 + \mu_t} \tag{9.37}$$

$$r_{c.v1.hi} = \frac{r_{a.t} + Z_{v2.hi}}{1 + \mu_t} \tag{9.38}$$

with V1 cathode output resistances $R_{c.v1.lo}$ and $R_{c.v1.hi}$:

$$R_{c.v1.lo} = r_{c.v1.lo} \parallel R_c$$
$$= \left(\frac{1}{r_{c.v1.lo}} + \frac{1}{R_c}\right)^{-1} \tag{9.39}$$

$$R_{c.v1.hi} = r_{c.v1.hi} \parallel R_c$$
$$= \left(\frac{1}{r_{c.v1.hi}} + \frac{1}{R_c}\right)^{-1} \tag{9.40}$$

capacitances $C_{c1.lo}$ and $C_{c1.hi}$ become ($f_{c.opt} \leq 0.2Hz$):

$$C_{c1.lo} = \frac{1}{2\pi f_{c.opt} R_{c.v1.lo}}$$ (9.41)

$$C_{c1.hi} = \frac{1}{2\pi f_{c.opt} R_{c.v1.hi}}$$ (9.42)

9.5.2.5   un-bypassed version -   V1 anode output resistance $R_{o.a.v1.u}$
                                      and impedance $Z_{o.a.v1.u}(f)$:

$$R_{o.a.v1.u} = r_{a.t} + (1+\mu_t) R_c$$ (9.43)

$$Z_{o.a.v1.u}(f) = R_{o.a.v1.u} \| C_{o.v1}$$ (9.44)

9.5.2.6   un-bypassed version -   AIK gain stage o/p resistances $R_{o.aik.u.lo}$ and $R_{o.aik.u.hi}$
                                      and o/p impedances $Z_{o.aik.u.lo}(f)$ and $Z_{o.aik.u.hi}(f)$ at the
                                      plate of V1:

$$R_{o.aik.u.lo} = R_{o.a.v1.u} \| R_{v2.lo}$$ (9.45)

$$R_{o.aik.u.hi} = R_{o.a.v1.u} \| R_{v2.hi}$$ (9.46)

$$Z_{o.aik.u.lo}(f) = Z_{o.a.v1.u}(f) \| Z_{v2.lo}(f)$$ (9.47)

$$Z_{o.aik.u.hi}(f) = Z_{o.a.v1.u}(f) \| Z_{v2.hi}(f)$$ (9.48)

9.6 Input capacitances $C_{i.v1}$ and $C_{i.v2}$
Output capacitances $C_{o.v1}$ & $C_{o.v2}$ and C_L

data sheet figures:

$C_{g.c.v1}$ = grid-cathode capacitance of V1
$C_{g.a.v1}$ = grid-plate capacitance of V1
$C_{a.c.v1}$ = plate-cathode capacitance of V1
$C_{g.c.v2}$ = grid-cathode capacitance of V2
$C_{g.a.v2}$ = grid-plate capacitance of V2
$C_{a.c.v2}$ = plate-cathode capacitance of V2

to be guessed:

$C_{stray.1}$ = sum of stray capacitances around V1
$C_{stray.2}$ = sum of stray capacitances around V2

Note: the Miller capacitance of the input depends on V1's gain G_{aik} only[10]!

$$C_{i.v1.b.lo.eff} = \left(1 + \left|G_{aik.b.lo.eff}\right|\right) C_{g.a.v1} \parallel C_{g.c.v1} \parallel C_{stray.1} \tag{9.49}$$

$$C_{i.v1.b.hi.eff} = \left(1 + \left|G_{aik.b.hi.eff}\right|\right) C_{g.a.v1} \parallel C_{g.c.v1} \parallel C_{stray.1} \tag{9.50}$$

$$C_{i.v1.u.lo.eff} = \left(1 + \left|G_{aik.u.lo.eff}\right|\right) C_{g.a.v1} \parallel C_{g.c.v1} \parallel C_{stray.1} \tag{9.51}$$

$$C_{i.v1.u.hi.eff} = \left(1 + \left|G_{aik.u.hi.eff}\right|\right) C_{g.a.v1} \parallel C_{g.c.v1} \parallel C_{stray.1} \tag{9.52}$$

$$C_{o.v1.lo} = C_{a.c.v1} \parallel C_{g.a.v1} \parallel C_{g.a.v2} \parallel C_{stray.2} \tag{9.53}$$

$$C_{o.v2.lo} = C_{a.c.v2} \tag{9.54}$$

$$C_{o.v1.hi} = C_{a.c.v1} \parallel C_{g.a.v1} \tag{9.55}$$

$$C_{o.v2.hi} = C_{a.c.v2} \parallel C_{stray.2} \tag{9.56}$$

output capacitance C_L:

to simulate a nearly real situation of an AIK gain stage with its rather high o/p impedance I've included a (tiny) load capacitance C_L into the calculation course, hence, Z_L becomes:

$$Z_L = R_L \parallel C_L \tag{9.57}$$

---

[10]   Miller capacitance effects on C_i and C_o: see Chapter 1, paragraph 1.3.2.5

9.7   Input impedance Z_{in}

$$Z_{in}(f) = R_{g1} \parallel C_{i.v1} \qquad (9.58)$$

Inclusion of R_{gg1}[11] and C_{in} into the calculation course leads to:

$$Z_{in.b.lo.eff}(f) = \left[\left(R_{gg1} + \frac{1}{2j\pi f\, C_{i.v1.b.lo.eff}} \right)^{-1} + \frac{1}{R_{g1}} \right]^{-1} + \frac{1}{2j\pi f\, C_{in}} \qquad (9.59)$$

$$Z_{in.b.hi.eff}(f) = \left[\left(R_{gg1} + \frac{1}{2j\pi f\, C_{i.v1.b.hi.eff}} \right)^{-1} + \frac{1}{R_{g1}} \right]^{-1} + \frac{1}{2j\pi f\, C_{in}} \qquad (9.60)$$

$$Z_{in.u.lo.eff}(f) = \left[\left(R_{gg1} + \frac{1}{2j\pi f\, C_{i.v1.u.lo.eff}} \right)^{-1} + \frac{1}{R_{g1}} \right]^{-1} + \frac{1}{2j\pi f\, C_{in}} \qquad (9.61)$$

$$Z_{in.u.hi.eff}(f) = \left[\left(R_{gg1} + \frac{1}{2j\pi f\, C_{i.v1.u.hi.eff}} \right)^{-1} + \frac{1}{R_{g1}} \right]^{-1} + \frac{1}{2j\pi f\, C_{in}} \qquad (9.62)$$

---

[11] R_{gg1} = oscillation prevention resistor between grid of V1 and R_{g1}

9.8   Gain stage frequency and phase response calculations[12]

9.8.1 Bypassed version:

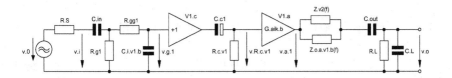

Figure 9.4 Simplified[13] equivalent circuit of Figure 9.1 (bypassed version)
- including all frequency and phase response relevant components

9.8.1.1 1st stage (V1 = $V1_c$+$V1_a$) input transfer function $T_{i.1.b}(f)$ and phase $\varphi_{i.1.b}(f)$ -
including source resistance R_S and an oscillation preventing resistor R_{gg1}<<R_{g1}:

$$T_{i.1.b}(f)=\frac{v_{g.1}}{v_0} \tag{9.63}$$

$$\varphi_{i.1.b}(f)=\arctan\left\{\frac{\operatorname{Im}\left[T_{i.1.b}(f)\right]}{\operatorname{Re}\left[T_{i.1.b}(f)\right]}\right\} \tag{9.64}$$

$$T_{i.1.b}(f)=\frac{Z2_b(f)\left(\dfrac{1}{R_{g1}}+\dfrac{1}{R_{gg1}+Z2_b(f)}\right)^{-1}}{\left(R_{gg1}+Z2_b(f)\right)\left[R_S+Z1(f)+\left(\dfrac{1}{R_{g1}}+\dfrac{1}{R_{gg1}+Z2_b(f)}\right)^{-1}\right]} \tag{9.65}$$

$$Z1(f)=\left(2j\pi f\,C_{in}\right)^{-1}$$
$$Z2_b(f)=\left(2j\pi f\,C_{i.v1.b}\right)^{-1} \tag{9.66}$$

9.8.1.2 1st stage $V1_c$ cathode transfer function $T_{c1.b}(f)$ and phase $\varphi_{c1.b}(f)$:

$$\varphi_{c1.b}(f)=\arctan\left\{\frac{\operatorname{Im}\left[T_{c1.b}(f)\right]}{\operatorname{Re}\left[T_{c1.b}(f)\right]}\right\} \tag{9.67}$$

---

[12]   By inclusion of the subscripts "lo" and "hi" the following formulae can be used for both the
lo and hi versions (see respective MCD worksheets IX-1 and IX-2)
[13]   See footnotes 2 and 3 of Chapter 1 and footnote 3 of Chapter 2

$$T_{c1.b}(f) = \frac{R_{c.v1}}{R_{c.v1} + \dfrac{1}{2j\pi f\, C_{c1}}} \tag{9.68}$$

9.8.1.3 1st stage output transfer function $T_{o.1.b}(f)$ and phase $\varphi_{o.1.b}(f)$[14]:

$$T_{o.1.b}(f) = \frac{v_o}{v_{a.1}} \tag{9.69}$$

$$\varphi_{o.1.b}(f) = \arctan\left\{\frac{\mathrm{Im}\left[T_{o.1.b}(f)\right]}{\mathrm{Re}\left[T_{o.1.b}(f)\right]}\right\} \tag{9.70}$$

$$T_{o.1.b}(f) = \frac{Z_L(f)}{Z_L(f) + Z3(f) + \left(\dfrac{1}{Z_{v2}(f)} + \dfrac{1}{Z_{o.a.v1.b}(f)}\right)^{-1}} \tag{9.71}$$

$$Z3(f) = \left(2j\pi f\, C_{out}\right)^{-1}$$

$$Z_L(f) = \left(\frac{1}{R_L} + 2j\pi f C_L\right)^{-1} \tag{9.72}$$

9.8.1.4 1st stage fundamental phase shift $\varphi_{G.v1}(f)$:

$$\varphi_{G.v1}(f) = \varphi_{G.aik.b}(f) = -180° \tag{9.73}$$

9.8.1.5   total gain stage transfer function $T_{tot.b}(f)$ and phase $\varphi_{tot.b}(f)$:

$$T_{tot.b}(f) = T_{i.1.b}(f)\, T_{c1.b}(f)\, T_{o.1.b}(f)\, G_{aik.b} \tag{9.74}$$

$$\varphi_{tot.b}(f) = \varphi_{i.1.b}(f) + \varphi_{c1.b}(f) + \varphi_{o.1.b}(f) + \varphi_{G.aik.b}(f) \tag{9.75}$$

---

[14] R_L and C_L need to be carefully chosen because they heavily influence the gain and phase of the AIK; R_L should be chosen as high as possible, eg. a CF2 à la Chapter 3

9.8.2 Un-bypassed version:

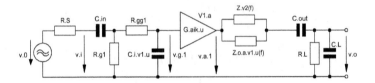

Figure 9.5 Simplified[15] equivalent circuit of Figure 9.1 (un-bypassed version)
- including all frequency and phase response relevant components

9.8.2.1   1st stage (V1$_a$) input transfer function $T_{i.1.u}(f)$ and phase $\varphi_{i.1.u}(f)$ - including source
resistance R_S and an oscillation preventing resistor $R_{gg1} \ll R_{g1}$:

$$T_{i.1.u}(f) = \frac{v_{g.1}}{v_0} \tag{9.76}$$

$$\varphi_{i.1.u}(f) = \arctan\left\{\frac{\text{Im}\left[T_{i.1.u}(f)\right]}{\text{Re}\left[T_{i.1.u}(f)\right]}\right\} \tag{9.77}$$

$$T_{i.1.u}(f) = \frac{Z2_u(f)\left(\dfrac{1}{R_{g1}} + \dfrac{1}{R_{gg1} + Z2_u(f)}\right)^{-1}}{\left(R_{gg1} + Z2_u(f)\right)\left[R_S + Z1(f) + \left(\dfrac{1}{R_{g1}} + \dfrac{1}{R_{gg1} + Z2_u(f)}\right)^{-1}\right]} \tag{9.78}$$

$$Z1(f) = \left(2j\pi f\, C_{in}\right)^{-1}$$
$$Z2_u(f) = \left(2j\pi f\, C_{i.v1.u}\right)^{-1} \tag{9.79}$$

9.8.2.2   1st stage output transfer function $T_{o.1.u}(f)$ and phase $\varphi_{o.1.u}(f)$[16]:

$$T_{o.1.u}(f) = \frac{v_o}{v_{a.1}} \tag{9.80}$$

$$\varphi_{o.1.u}(f) = \arctan\left\{\frac{\text{Im}\left[T_{o.1.u}(f)\right]}{\text{Re}\left[T_{o.1.u}(f)\right]}\right\} \tag{9.81}$$

---

[15]   See also footnotes 2 and 3 of Chapter 1
[16]   see footnote 7

$$T_{o.1.u}(f) = \frac{Z_L(f)}{Z_L(f) + Z3(f) + \left(\dfrac{1}{Z_{v2}(f)} + \dfrac{1}{Z_{o.a.v1.u}(f)}\right)^{-1}} \qquad (9.82)$$

$$Z3(f) = (2j\pi f\, C_{out})^{-1}$$

$$Z_L(f) = \left(\frac{1}{R_L} + 2j\pi f C_L\right)^{-1} \qquad (9.83)$$

9.8.2.3 1st stage fundamental phase shift $\varphi_{G.v1}(f)$:

$$\varphi_{G.v1}(f) = \varphi_{G.aik.u}(f) = -180° \qquad (9.84)$$

9.8.2.4 total gain stage transfer function $T_{tot.u}(f)$ and phase $\varphi_{tot.u}(f)$:

$$T_{tot.u}(f) = T_{i.1.u}(f)\, T_{o.1.u}(f)\, G_{aik.u} \qquad (9.85)$$

$$\varphi_{tot.u}(f) = \varphi_{i.1.u}(f) + \varphi_{o.1.u}(f) + \varphi_{G.aik.u}(f) \qquad (9.86)$$

9.9   Example with ECC83 / 12AX7 (83 & lo):

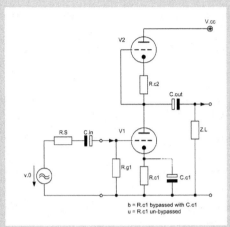

Figure 9.6   AIK example circuitry with a low-Z current source plate load

9.9.1   Triode bias data:

$$I_{a1.83} = I_{a2.83} = 1.2 \cdot 10^{-3} A \qquad V_{a1.83} = 100V \qquad V_{a2.83} = 100V$$

$$V_{cc} = 201V \qquad V_{g1.83} = -0.5V \qquad V_{g2.83} = 100.5V$$

9.9.2   Triode valve constants:

$$g_{m.83} := 1.75 \cdot 10^{-3} \cdot S \qquad \mu_{83} := 101.5 \qquad r_{a.83} := 58 \cdot 10^{3} \Omega$$

$$g_{m.v1} := g_{m.83} \qquad \mu_{v1} := \mu_{83} \qquad r_{a.v1} := r_{a.83}$$

$$g_{m.v2} := g_{m.83} \qquad \mu_{v2} := \mu_{83} \qquad r_{a.v2} := r_{a.83}$$

$$C_{g.c.1.83} := 1.65 \cdot 10^{-12} F \qquad C_{g.a.1.83} := 1.6 \cdot 10^{-12} F \qquad C_{a.c.1.83} := 0.33 \cdot 10^{-12} F$$

$$C_{g.c.2.83} := 1.65 \cdot 10^{-12} F \qquad C_{g.a.2.83} := 1.6 \cdot 10^{-12} F \qquad C_{a.c.2.83} := 0.33 \cdot 10^{-12} F$$

9.9.3   Circuit variables:

$$R_{c1.83} := 417\Omega \qquad R_{c2.83} := R_{c1.83} \qquad R_{c.83} := R_{c1.83}$$

$$R_{S} := 1 \cdot 10^{3} \Omega \qquad R_{g1.83} := 47.5 \cdot 10^{3} \Omega \qquad R_{gg1.83} := 0.22 \cdot 10^{3} \Omega$$

$$R_{L} := 1 \cdot 10^{6} \Omega \qquad C_{in.83} := 2.2 \cdot 10^{-6} F \qquad C_{out.83} := 1 \cdot 10^{-6} F$$

$$C_{L} := 10 \cdot 10^{-12} F \qquad C_{stray.83.1} := 10 \cdot 10^{-12} F \qquad C_{stray.83.2} := 2 \cdot 10^{-12} F$$

➢   MCD Worksheet IX-1   AIK -lo- calculations   page 2

9.9.4   Calculation relevant data:

frequency range f for the below shown graphs: $f := 10\,Hz, 20\,Hz .. 100000\,Hz$

$h := 1000 \cdot Hz$

9.9.5   Gain $G_{aik.83.b.lo}$:

$R_{v2.lo} := r_{a.83} + \left(1 + \mu_{83}\right) \cdot R_{c1.83}$ $R_{v2.lo} = 100.743 \times 10^3\,\Omega$

$G_{aik.83.b.lo} := -g_{m.83} \cdot \left(\dfrac{1}{r_{a.83}} + \dfrac{1}{R_{v2.lo}}\right)^{-1}$ $G_{aik.83.b.lo} = -64.415 \times 10^0$

$G_{aik.83.b.lo.e} := 20 \cdot \log\left(\left|G_{aik.83.b.lo}\right|\right)$ $G_{aik.83.b.lo.e} = 36.18 \times 10^0$ [dB]

$G_{aik.83.b.lo.eff} := -g_{m.83} \cdot \left(\dfrac{1}{r_{a.83}} + \dfrac{1}{R_{v2.lo}} + \dfrac{1}{R_L}\right)^{-1}$ $G_{aik.83.b.lo.eff} = -62.128 \times 10^0$

$G_{aik.83.b.lo.eff.e} := 20 \cdot \log\left(\left|G_{aik.83.b.lo.eff}\right|\right)$ $G_{aik.83.b.lo.eff.e} = 35.866 \times 10^0$ [dB]

9.9.6   Gain $G_{aik.83.u.lo}$:

$G_{aik.83.u.lo} := -\dfrac{1}{2} \cdot g_{m.83} \cdot r_{a.83}$ $G_{aik.83.u.lo} = -50.75 \times 10^0$

$G_{aik.83.u.lo.e} := 20 \cdot \log\left(\left|G_{aik.83.u.lo}\right|\right)$ $G_{aik.83.u.lo.e} = 34.109 \times 10^0$ [dB]

$G_{aik.83.u.lo.eff} := -g_{m.83} \cdot \dfrac{r_{a.83} \cdot R_L}{r_{a.83} + \left(1 + \mu_{83}\right) \cdot R_{c.83} + 2 \cdot R_L}$ $G_{aik.83.u.lo.eff} = -48.316 \times 10^0$

$G_{aik.83.u.lo.eff.e} := 20 \cdot \log\left(\left|G_{aik.83.u.lo.eff}\right|\right)$ $G_{aik.83.u.lo.eff.e} = 33.682 \times 10^0$ [dB]

9.9.7   Specific resistances:

$R_{o.a.v1.b} := r_{a.v1}$ $R_{o.a.v1.b} = 58 \times 10^3\,\Omega$

$R_{o.aik.b.lo} := \left(\dfrac{1}{R_{o.a.v1.b}} + \dfrac{1}{R_{v2.lo}}\right)^{-1}$ $R_{o.aik.b.lo} = 36.808 \times 10^3\,\Omega$

$r_{c.v1.lo} := \dfrac{r_{a.v1} + R_{v2.lo}}{1 + \mu_{v1}}$ $r_{c.v1.lo} = 1.549 \times 10^3\,\Omega$

$R_{c.v1.lo} := \left(\dfrac{1}{r_{c.v1.lo}} + \dfrac{1}{R_{c1.83}}\right)^{-1}$ $R_{c.v1.lo} = 328.539 \times 10^0\,\Omega$

$R_{o.a.v1.u} := r_{a.v1} + \left(1 + \mu_{v1}\right) \cdot R_{c1.83}$ $R_{o.a.v1.u} = 100.743 \times 10^3\,\Omega$

$R_{o.aik.u.lo} := \left(\dfrac{1}{R_{o.a.v1.u}} + \dfrac{1}{R_{v2.lo}}\right)^{-1}$ $R_{o.aik.u.lo} = 50.371 \times 10^3\,\Omega$

9.9.8 R_L variations:

$$R_{L.var} := 10 \cdot 10^3 \Omega, 20 \cdot 10^3 \Omega .. 10 \cdot 10^6 \Omega$$

$$G_{aik.b.lo.eff}(R_{L.var}) := -g_{m.83} \cdot \left(\frac{1}{r_{a.83}} + \frac{1}{R_{v2.lo}} + \frac{1}{R_{L.var}} \right)^{-1}$$

$$G_{aik.u.lo.eff}(R_{L.var}) := -g_{m.83} \cdot \frac{r_{a.83} \cdot R_{L.var}}{r_{a.83} + (1 + \mu_{83}) \cdot R_{c.83} + 2 \cdot R_{L.var}}$$

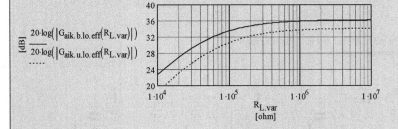

$$\begin{array}{l} \underline{\quad} \quad 20 \cdot \log\left(\left| G_{aik.b.lo.eff}(R_{L.var}) \right| \right) \\ \\ \underline{\quad} \quad 20 \cdot \log\left(\left| G_{aik.u.lo.eff}(R_{L.var}) \right| \right) \\ \text{-----} \end{array}$$

Figure 9.7 Effecive gains vs. R_L changes

9.9.9   Capacitance C_{c1}:

$$f_c := 20 Hz \qquad\qquad f_{c.opt} := \frac{f_c}{100} \qquad\qquad f_{c.opt} = 200 \times 10^{-3} Hz$$

$$C_{c1.lo} := \frac{1}{2 \cdot \pi \cdot f_{c.opt} \cdot R_{c.v1.lo}} \qquad\qquad C_{c1.lo} = 2.422 \times 10^{-3} F$$

9.9.10   Specific input and output capacitances:

$$C_{i.v1.b.lo} := \left(1 + \left| G_{aik.83.b.lo} \right| \right) \cdot C_{g.a.1.83} + C_{g.c.1.83} + C_{stray.83.1} \qquad C_{i.v1.b.lo} = 116.314 \times 10^{-12} F$$

$$C_{i.v1.u.lo} := \left(1 + \left| G_{aik.83.u.lo} \right| \right) \cdot C_{g.a.1.83} + C_{g.c.1.83} + C_{stray.83.1} \qquad C_{i.v1.u.lo} = 94.45 \times 10^{-12} F$$

$$C_{o.v1.lo} := C_{a.c.1.83} + C_{g.a.1.83} + C_{g.a.2.83} + C_{stray.83.2} \qquad C_{o.v1.lo} = 5.53 \times 10^{-12} F$$

$$C_{o.v2.lo} := C_{a.c.2.83} \qquad C_{o.v2.lo} = 330 \times 10^{-15} F$$

9.9.11   Specific impedances:

$$X_{c.2.83}(f) := (1 + \mu_{83}) \cdot \left(\frac{1}{R_{c2.83}} + 2j \cdot \pi \cdot f \cdot C_{g.c.2.83} \right)^{-1} \qquad Y_{c.2.83}(f) := \left(\frac{1}{r_{a.v2}} + 2j \cdot \pi \cdot f \cdot C_{o.v2.lo} \right)^{-1}$$

$$Z_{v2.lo}(f) := \left[\left(Y_{c.2.83}(f) + X_{c.2.83}(f) \right)^{-1} + 2j \cdot \pi \cdot f \cdot \left(C_{g.a.2.83} + C_{stray.83.2} \right) \right]^{-1}$$

$$\left| Z_{v2.lo}(h) \right| = 100.742 \times 10^3 \Omega \qquad\qquad \left| Z_{v2.lo}(20 \cdot 10^3 Hz) \right| = 100.631 \times 10^3 \Omega$$

➢ MCD Worksheet IX-1   AIK -lo- calculations   page 4

$$Z_{o.a.vl.b}(f) := \left(\frac{1}{R_{o.a.vl.b}} + 2j \cdot \pi \cdot f \cdot C_{o.vl.lo} \right)^{-1}$$

$$\left| Z_{o.a.vl.b}(h) \right| = 58 \times 10^3 \, \Omega$$

$$Z_{o.aik.b.lo}(f) := \left(\frac{1}{Z_{o.a.vl.b}(f)} + \frac{1}{Z_{v2.lo}(f)} \right)^{-1}$$

$$\left| Z_{o.aik.b.lo}(h) \right| = 36.808 \times 10^3 \, \Omega$$

$$Z_{o.a.vl.u}(f) := \left(\frac{1}{R_{o.a.vl.u}} + 2j \cdot \pi \cdot f \cdot C_{o.vl.lo} \right)^{-1}$$

$$\left| Z_{o.a.vl.u}(h) \right| = 100.742 \times 10^3 \, \Omega$$

$$Z_{o.aik.u.lo}(f) := \left(\frac{1}{Z_{o.a.vl.u}(f)} + \frac{1}{Z_{v2.lo}(f)} \right)^{-1}$$

$$\left| Z_{o.aik.u.lo}(h) \right| = 50.371 \times 10^3 \, \Omega$$

9.9.12 Gain stage frequency and phase response
- bypassed version:

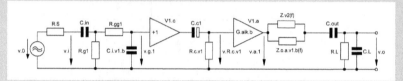

Figure 9.8 = Figure 9.4

$$Z1(f) := \frac{1}{2j \cdot \pi \cdot f \cdot C_{in.83}}$$

$$Z2_{b.lo}(f) := \frac{1}{2j \cdot \pi \cdot f \cdot C_{i.vl.b.lo}}$$

$$T_{i.1.b.lo}(f) := \frac{Z2_{b.lo}(f) \cdot \left(\frac{1}{R_{g1.83}} + \frac{1}{R_{gg1.83} + Z2_{b.lo}(f)} \right)^{-1}}{\left(Z2_{b.lo}(f) + R_{gg1.83} \right) \cdot \left[R_S + Z1(f) + \left(\frac{1}{R_{g1.83}} + \frac{1}{R_{gg1.83} + Z2_{b.lo}(f)} \right)^{-1} \right]}$$

$$T_{i.1.b.lo.e}(f) := 20 \cdot \log \left(\left| T_{i.1.b.lo}(f) \right| \right)$$

$$\phi_{i.1.b.lo}(f) := \operatorname{atan}\left(\frac{\operatorname{Im}\left(T_{i.1.b.lo}(f) \right)}{\operatorname{Re}\left(T_{i.1.b.lo}(f) \right)} \right)$$

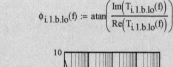

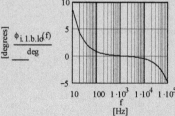

Figure 9.9 Transfer of i/p network   Figure 9.10 Phase of i/p network

➢   MCD Worksheet IX-1   AIK -lo- calculations   page 5

$$T_{c1.b.lo}(f) := \left(\frac{R_{c.v1.lo}}{R_{c.v1.lo} + \dfrac{1}{2j \cdot \pi \cdot f \cdot C_{c1.lo}}} \right)$$

$$\phi_{c1.b.lo}(f) := atan\left(\frac{Im(T_{c1.b.lo}(f))}{Re(T_{c1.b.lo}(f))} \right)$$

$$T_{c1.b.lo.e}(f) := 20 \cdot log(|T_{c1.b.lo}(f)|)$$

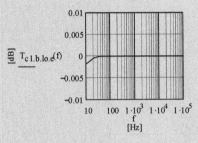

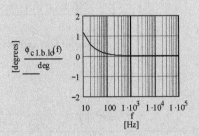

Figure 9.11 Transfer of V1 cathode network   Figure 9.12 Phase of V1 cathode network

$$Z4(f) := \left(\frac{1}{R_L} + 2j \cdot \pi \cdot f \cdot C_L \right)^{-1}$$

$$Z3(f) := \left(2j \cdot \pi \cdot f \cdot C_{out.83} \right)^{-1}$$

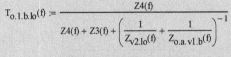
$$T_{o.1.b.lo}(f) := \frac{Z4(f)}{Z4(f) + Z3(f) + \left(\dfrac{1}{Z_{v2.lo}(f)} + \dfrac{1}{Z_{o.a.v1.b}(f)} \right)^{-1}}$$

$$\phi_{o.1.b.lo}(f) := atan\left(\frac{Im(T_{o.1.b.lo}(f))}{Re(T_{o.1.b.lo}(f))} \right)$$

$$T_{o.1.b.lo.e}(f) := 20 \cdot log(|T_{o.1.b.lo}(f)|)$$

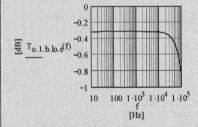

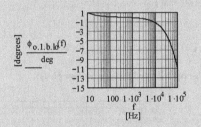

Figure 9.13 Transfer of V1 o/p network   Figure 9.14 Phase of V1 o/p network

$$T_{tot.83.b.lo}(f) := T_{i.1.b.lo}(f) \cdot T_{c1.b.lo}(f) \cdot T_{o.1.b.lo}(f) \cdot G_{aik.83.b.lo}$$

$$\phi_{G.v1}(f) := -180 \, deg$$

$$T_{tot.83.b.lo.e}(f) := 20 \cdot log(|T_{tot.83.b.lo}(f)|)$$

$$T_{tot.83.b.lo.e}(h) = 35.684 \times 10^{0} \ [dB]$$

$$\phi_{tot.83.b.lo}(f) := \phi_{i.1.b.lo}(f) + \phi_{c1.b.lo}(f) + \phi_{o.1.b.lo}(f) + \phi_{G.v1}(f)$$

➤ MCD Worksheet IX-1   AIK -lo- calculations   page 6

9.9.13 Gain stage frequency and phase response
- un-bypassed version:

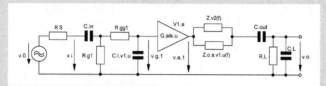

Figure 9.15 = Figure 9.5

$$Z2_{u.lo}(f) := \frac{1}{2j \cdot \pi \cdot f \cdot C_{i.v1.u.lo}}$$

$$T_{i.1.u.lo}(f) := \frac{Z2_{u.lo}(f) \cdot \left(\dfrac{1}{R_{g1.83}} + \dfrac{1}{R_{gg1.83} + Z2_{u.lo}(f)} \right)^{-1}}{\left(Z2_{u.lo}(f) + R_{gg1.83} \right) \left[R_S + Z1(f) + \left(\dfrac{1}{R_{g1.83}} + \dfrac{1}{R_{gg1.83} + Z2_{u.lo}(f)} \right)^{-1} \right]}$$

$$\phi_{i.1.u.lo}(f) := atan \left(\frac{Im\left(T_{i.1.u.lo}(f) \right)}{Re\left(T_{i.1.u.lo}(f) \right)} \right)$$

$$T_{o.1.u.lo}(f) := \frac{Z4(f)}{Z4(f) + Z3(f) + \left(\dfrac{1}{Z_{v2.lo}(f)} + \dfrac{1}{Z_{o.a.v1.u}(f)} \right)^{-1}}$$

$$\phi_{o.1.u.lo}(f) := atan \left(\frac{Im\left(T_{o.1.u.lo}(f) \right)}{Re\left(T_{o.1.u.lo}(f) \right)} \right)$$

$$T_{tot.83.u.lo}(f) := T_{i.1.u.lo}(f) \, T_{o.1.u.lo}(f) \cdot G_{aik.83.u.lo}$$

$$T_{tot.83.u.lo.e}(f) := 20 \cdot log\left(\left| T_{tot.83.u.lo}(f) \right| \right) \qquad\qquad T_{tot.83.u.lo.e}(h) = 33.5 \times 10^0 \qquad [dB]$$

$$\phi_{tot.83.u.lo}(f) := \phi_{i.1.u.lo}(f) + \phi_{o.1.u.lo}(f) + \phi_{G.v1}(f)$$

9.9.14 Frequency and phase response plots:

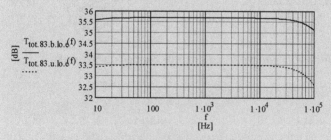

Figure 9.16 Frequency responses of the bypassed and un-bypassed
AIK gain stage

➤   MCD Worksheet IX-1   AIK -lo- calculations   page 7

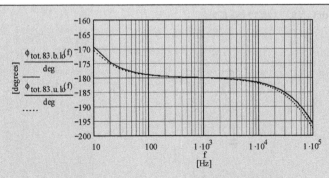

Figure 9.17 Phase responses of the bypassed and un-bypassed
AIK gain stage

9.9.15 Input impedances:

$$Z_{in.b.lo}(f) := \left[\left(R_{gg1.83} + \frac{1}{2j \cdot \pi \cdot f \cdot C_{i.v1.b.lo}}\right)^{-1} + \frac{1}{R_{g1.83}}\right]^{-1} + \frac{1}{2j \cdot \pi \cdot f \cdot C_{in.83}}$$

$$Z_{in.u.lo}(f) := \left[\left(R_{gg1.83} + \frac{1}{2j \cdot \pi \cdot f \cdot C_{i.v1.u.lo}}\right)^{-1} + \frac{1}{R_{g1.83}}\right]^{-1} + \frac{1}{2j \cdot \pi \cdot f \cdot C_{in.83}}$$

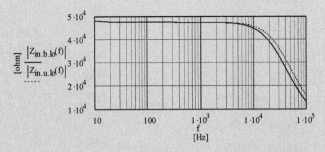

Figure 9.18 Input impedances for the bypassed and un-bypassed versions

9.10   Example with ECC83 / 12AX7 (83 & hi):

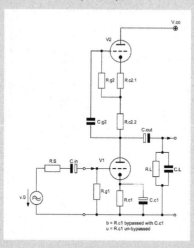

Figure 9.19 AIK example circuitry with a high-Z current source plate load

9.10.1 Triode bias data:

$$I_{a1.83} = I_{a2.83} = 1.2 \cdot 10^{-3}\,A \qquad V_{a1.83} = 100\,V \qquad\qquad V_{a2.83} = 100\,V$$

$$V_{cc} = 268.5\,V \qquad\qquad V_{g1.83} = -0.5\,V \qquad\qquad V_{g2.83} = 168\,V$$

9.10.2 Triode valve constants:

$$g_{m.83} := 1.75 \cdot 10^{-3} \cdot S \qquad \mu_{83} := 101.5 \qquad\qquad r_{a.83} := 58 \cdot 10^{3}\,\Omega$$

$$g_{m.v1} := g_{m.83} \qquad\qquad \mu_{v1} := \mu_{83} \qquad\qquad r_{a.v1} := r_{a.83}$$

$$g_{m.v2} := g_{m.83} \qquad\qquad \mu_{v2} := \mu_{83} \qquad\qquad r_{a.v2} := r_{a.83}$$

$$C_{g.c.1.83} := 1.65 \cdot 10^{-12}\,F \qquad C_{g.a.1.83} := 1.6 \cdot 10^{-12}\,F \qquad C_{a.c.1.83} := 0.33 \cdot 10^{-12}\,F$$

$$C_{g.c.2.83} := 1.65 \cdot 10^{-12}\,F \qquad C_{g.a.2.83} := 1.6 \cdot 10^{-12}\,F \qquad C_{a.c.2.83} := 0.33 \cdot 10^{-12}\,F$$

9.10.3 Circuit variables:

$$R_{c1.83} := 417\,\Omega \qquad\qquad R_{c2.1.83} := R_{c1.83} \qquad\qquad R_{c2.2.83} := 56.2 \cdot 10^{3}\,\Omega$$

$$R_{S} := 1 \cdot 10^{3}\,\Omega \qquad\qquad R_{g1.83} := 47.5 \cdot 10^{3}\,\Omega \qquad R_{gg1.83} := 0.22 \cdot 10^{3}\,\Omega$$

$$R_{L} := 5 \cdot 10^{6}\,\Omega \qquad\qquad R_{g2.83} := 1 \cdot 10^{6}\,\Omega$$

$$C_{L} := 2 \cdot 10^{-12}\,F \qquad\qquad C_{stray.83.1} := 10 \cdot 10^{-12}\,F \qquad C_{stray.83.2} := 2 \cdot 10^{-12}\,F$$

$$C_{in.83} := 2.2 \cdot 10^{-6}\,F \qquad\qquad C_{out.83} := 1 \cdot 10^{-6}\,F$$

➢   MCD Worksheet IX-2   AIK -hi- calculations   page 2

9.10.4 Calculation relevant data:

frequency range f for the below shown graphs: $f := 10Hz, 20Hz .. 100000\,Hz$

$h := 1000 \cdot Hz$

9.10.5 Gain $G_{aik.83.b.hi}$:

$R_{p2.83} := \left(\dfrac{1}{R_{c2.2.83}} + \dfrac{1}{R_{g2.83}} \right)^{-1}$

$R_{v2.hi} := r_{a.83} + \left(1 + \mu_{83} \right) \cdot \left(R_{c2.1.83} + R_{p2.83} \right)$

$R_{p3.83} := \left(\dfrac{1}{R_{v2.hi}} + \dfrac{1}{R_L} \right)^{-1}$

$R_{v2.hi} = 5.555 \times 10^6 \, \Omega$

$G_{aik.83.b.hi} := -g_{m.83} \cdot \left(\dfrac{1}{r_{a.83}} + \dfrac{1}{R_{v2.hi}} \right)^{-1}$

$G_{aik.83.b.hi} = -100.451 \times 10^0$

$G_{aik.83.b.hi.e} := 20 \cdot log \left(\left| G_{aik.83.b.hi} \right| \right)$

$G_{aik.83.b.hi.e} = 40.039 \times 10^0$ [dB]

$G_{aik.83.b.hi.eff} := -g_{m.83} \cdot \left(\dfrac{1}{r_{a.83}} + \dfrac{1}{R_{p3.83}} \right)^{-1}$

$G_{aik.83.b.hi.eff} = -99.311 \times 10^0$

$G_{aik.83.b.hi.eff.e} := 20 \cdot log \left(\left| G_{aik.83.b.hi.eff} \right| \right)$

$G_{aik.83.b.hi.eff.e} = 39.94 \times 10^0$ [dB]

9.10.6 Gain $G_{aik.83.u.hi}$:

$G_{aik.83.u.hi} := -g_{m.83} \cdot \dfrac{r_{a.83} \cdot R_{v2.hi}}{r_{a.83} + \left(1 + \mu_{83} \right) \cdot R_{c1.83} + R_{v2.hi}}$

$G_{aik.83.u.hi} = -99.692 \times 10^0$

$G_{aik.83.u.hi.e} := 20 \cdot log \left(\left| G_{aik.83.u.hi} \right| \right)$

$G_{aik.83.u.hi.e} = 39.973 \times 10^0$ [dB]

$G_{aik.83.u.hi.eff} := -g_{m.83} \cdot \dfrac{r_{a.83} \cdot R_{p3.83}}{r_{a.83} + \left(1 + \mu_{83} \right) \cdot R_{c1.83} + R_{p3.83}}$

$G_{aik.83.u.hi.eff} = -97.757 \times 10^0$

$G_{aik.83.u.hi.eff.e} := 20 \cdot log \left(\left| G_{aik.83.u.hi.eff} \right| \right)$

$G_{aik.83.u.hi.eff.e} = 39.803 \times 10^0$ [dB]

➢   **MCD Worksheet IX-2   AIK -hi- calculations   page 3**

9.10.7 Specific resistances:

$R_{o.a.vl.b} := r_{a.vl}$ $R_{o.a.vl.b} = 58 \times 10^3\, \Omega$

$R_{o.aik.b.hi} := \left(\dfrac{1}{R_{o.a.vl.b}} + \dfrac{1}{R_{v2.hi}} \right)^{-1}$ $R_{o.aik.b.hi} = 57.401 \times 10^3\, \Omega$

$r_{c.vl.hi} := \dfrac{r_{a.vl} + R_{v2.hi}}{1 + \mu_{vl}}$ $r_{c.vl.hi} = 54.758 \times 10^3\, \Omega$

$R_{c.vl.hi} := \left(\dfrac{1}{r_{c.vl.hi}} + \dfrac{1}{R_{c1.83}} \right)^{-1}$ $R_{c.vl.hi} = 413.848 \times 10^0\, \Omega$

$R_{o.a.vl.u} := r_{a.vl} + \left(1 + \mu_{vl}\right) \cdot R_{c1.83}$ $R_{o.a.vl.u} = 100.743 \times 10^3\, \Omega$

$R_{o.aik.u.hi} := \left(\dfrac{1}{R_{o.a.vl.u}} + \dfrac{1}{R_{v2.hi}} \right)^{-1}$ $R_{o.aik.u.hi} = 98.948 \times 10^3\, \Omega$

9.10.8 R_L variations:

$R_{L.var} := 10 \cdot 10^3\, \Omega, 20 \cdot 10^3\, \Omega \,..\, 10 \cdot 10^6\, \Omega$

$G_{aik.b.hi.eff}\!\left(R_{L.var}\right) := -g_{m.83} \left(\dfrac{1}{r_{a.83}} + \dfrac{1}{R_{v2.hi}} + \dfrac{1}{R_{L.var}} \right)^{-1}$

$G_{aik.u.hi.eff}\!\left(R_{L.var}\right) := -g_{m.83} \cdot \dfrac{r_{a.83} \cdot R_{L.var}}{r_{a.83} + \left(1 + \mu_{83}\right) \cdot R_{c1.83} + 2 \cdot R_{L.var}}$

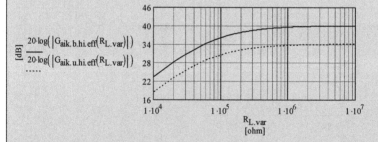

Figure 9.20 Effecive gains vs. R_L changes

9.10.9 Capacitance C_{c1}:

$f_c := 20\,\text{Hz}$ $f_{c.opt} := \dfrac{f_c}{100}$ $f_{c.opt} = 200 \times 10^{-3}\,\text{Hz}$

$C_{c1.hi} := \dfrac{1}{2 \cdot \pi \cdot f_{c.opt} \cdot R_{c.vl.hi}}$ $C_{c1.hi} = 1.923 \times 10^{-3}\,\text{F}$

9.10.10 Specific input and output capacitances:

$$C_{i.v1.b.hi} := \left(1 + |G_{aik.83.b.hi}|\right) \cdot C_{g.a.1.83} + C_{g.c.1.83} + C_{stray.83.1} \qquad C_{i.v1.b.hi} = 173.972 \times 10^{-12} F$$

$$C_{i.v1.u.hi} := \left(1 + |G_{aik.83.u.hi}|\right) \cdot C_{g.a.1.83} + C_{g.c.1.83} + C_{stray.83.1} \qquad C_{i.v1.u.hi} = 172.757 \times 10^{-12} F$$

$$C_{o.v1.hi} := C_{a.c.1.83} + C_{g.a.1.83} \qquad\qquad\qquad C_{o.v1.hi} = 1.93 \times 10^{-12} F$$

$$C_{o.v2.hi} := C_{a.c.2.83} + C_{stray.83.2} \qquad\qquad\qquad C_{o.v2.hi} = 2.33 \times 10^{-12} F$$

9.10.11 Specific impedances:

$$X_{v2.hi}(f) := \left(\frac{1}{R_{c2.1.83} + R_{p2.83}} + 2j \cdot \pi \cdot f C_{g.c.2.83}\right)^{-1} \qquad Y_{v2.hi}(f) := 2j \cdot \pi \cdot f \left(C_{g.a.2.83} + C_{stray.83.2}\right)$$

$$Z_{v2.hi}(f) := \left[\left[\left(\frac{1}{r_{a.v2}} + 2j \cdot \pi \cdot f C_{a.c.2.83}\right)^{-1} + \left(1 + \mu_{83}\right) \cdot X_{v2.hi}(f)\right]^{-1} + Y_{v2.hi}(f)\right]^{-1}$$

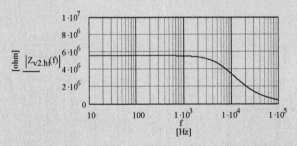

Figure 9.21 Hi-current source impedance vs. frequency

$$\left|Z_{v2.hi}(h)\right| = 5.511 \times 10^{6} \Omega$$

$$Z_{o.a.v1.b}(f) := \left(\frac{1}{R_{o.a.v1.b}} + 2j \cdot \pi \cdot f C_{o.v1.hi}\right)^{-1} \qquad \left|Z_{o.a.v1.b}(h)\right| = 58 \times 10^{3} \Omega$$

$$Z_{o.aik.b.hi}(f) := \left(\frac{1}{Z_{o.a.v1.b}(f)} + \frac{1}{Z_{v2.hi}(f)}\right)^{-1} \qquad \left|Z_{o.aik.b.hi}(h)\right| = 57.401 \times 10^{3} \Omega$$

$$Z_{o.a.v1.u}(f) := \left(\frac{1}{R_{o.a.v1.u}} + 2j \cdot \pi \cdot f C_{o.v1.hi}\right)^{-1} \qquad \left|Z_{o.a.v1.u}(h)\right| = 100.742 \times 10^{3} \Omega$$

$$Z_{o.aik.u.hi}(f) := \left(\frac{1}{Z_{o.a.v1.u}(f)} + \frac{1}{Z_{v2.hi}(f)}\right)^{-1} \qquad \left|Z_{o.aik.u.hi}(h)\right| = 98.947 \times 10^{3} \Omega$$

9.10.12 Gain stage frequency and phase response
- bypassed version:

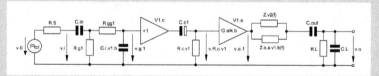

Figure 9.22 = Figure 9.4

$$Z1(f) := \frac{1}{2j \cdot \pi \cdot f \cdot C_{in.83}} \qquad\qquad Z2_{b.hi}(f) := \frac{1}{2j \cdot \pi \cdot f \cdot C_{i.v1.b.hi}}$$

$$T_{i.1.b.hi}(f) := \frac{Z2_{b.hi}(f) \cdot \left(\dfrac{1}{R_{g1.83}} + \dfrac{1}{R_{gg1.83} + Z2_{b.hi}(f)} \right)^{-1}}{\left(Z2_{b.hi}(f) + R_{gg1.83} \right) \cdot \left[R_S + Z1(f) + \left(\dfrac{1}{R_{g1.83}} + \dfrac{1}{R_{gg1.83} + Z2_{b.hi}(f)} \right)^{-1} \right]}$$

$$T_{i.1.b.hi.e}(f) := 20 \cdot \log\left(\left| T_{i.1.b.hi}(f) \right| \right) \qquad\qquad \phi_{i.1.b.hi}(f) := atan\left(\frac{Im\left(T_{i.1.b.hi}(f) \right)}{Re\left(T_{i.1.b.hi}(f) \right)} \right)$$

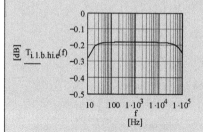

Figure 9.23 Transfer of i/p network   Figure 9.24 Phase of i/p network

$$T_{c1.b.hi}(f) := \left(\frac{R_{c.v1.hi}}{R_{c.v1.hi} + \dfrac{1}{2j \cdot \pi \cdot f \cdot C_{c1.hi}}} \right) \qquad \phi_{c1.b.hi}(f) := atan\left(\frac{Im\left(T_{c1.b.hi}(f) \right)}{Re\left(T_{c1.b.hi}(f) \right)} \right)$$

$$T_{c1.b.hi.e}(f) := 20 \cdot \log\left(\left| T_{c1.b.hi}(f) \right| \right)$$

> MCD Worksheet IX-2   AIK -hi- calculations   page 6

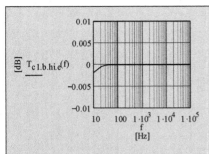

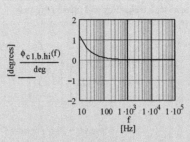

Figure 9.25 Transfer of V1 cathode network   Figure 9.26 Phase of V1 cathode network

$$Z3(f) := \frac{1}{2j \cdot \pi \cdot f \cdot C_{out.83}}$$

$$Z4(f) := \left(\frac{1}{R_L} + 2j \cdot \pi \cdot f \cdot C_L \right)^{-1}$$

$$T_{o.1.b.hi}(f) := \frac{Z4(f)}{Z4(f) + Z3(f) + \left(\frac{1}{Z_{v2.hi}(f)} + \frac{1}{Z_{o.a.v1.b}(f)} \right)^{-1}}$$

$$T_{o.1.b.hi.e}(f) := 20 \cdot \log \left(\left| T_{o.1.b.hi}(f) \right| \right)$$

$$\phi_{o.1.b.hi}(f) := atan\left(\frac{Im\left(T_{o.1.b.hi}(f) \right)}{Re\left(T_{o.1.b.hi}(f) \right)} \right)$$

$$T_{o.1.b.hi.e}(h) = -99.178 \times 10^{-3}$$

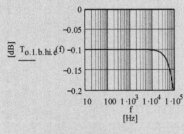

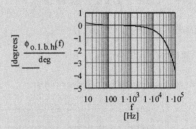

Figure 9.27 Transfer of V1 o/p network   Figure 9.28 Phase of V1 o/p network

$$T_{tot.83.b.hi}(f) := T_{i.1.b.hi}(f) \cdot T_{c1.b.hi}(f) \cdot T_{o.1.b.hi}(f) \cdot G_{aik.83.b.hi}$$

$$\phi_{G.v1}(f) := -180 deg$$

$$T_{tot.83.b.hi.e}(f) := 20 \cdot \log \left(\left| T_{tot.83.b.hi}(f) \right| \right)$$

$$T_{tot.83.b.hi.e}(h) = 39.758 \times 10^{0} \quad [dB]$$

$$\phi_{tot.83.b.hi}(f) := \phi_{i.1.b.hi}(f) + \phi_{c1.b.hi}(f) + \phi_{o.1.b.hi}(f) + \phi_{G.v1}(f)$$

> MCD Worksheet IX-2   AIK -hi- calculations   page 7

9.10.13 Gain stage frequency and phase response
- un-bypassed version:

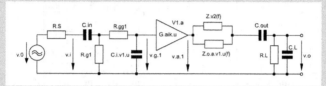

Figure 9.29 = Figure 9.5

$$Z2_{u.hi}(f) := \frac{1}{2j \cdot \pi \cdot f \cdot C_{i.v1.u.hi}}$$

$$T_{i.1.u.hi}(f) := \frac{Z2_{u.hi}(f) \cdot \left(\dfrac{1}{R_{g1.83}} + \dfrac{1}{R_{gg1.83} + Z2_{u.hi}(f)} \right)^{-1}}{\left(Z2_{u.hi}(f) + R_{gg1.83} \right) \cdot \left[R_S + Z1(f) + \left(\dfrac{1}{R_{g1.83}} + \dfrac{1}{R_{gg1.83} + Z2_{u.hi}(f)} \right)^{-1} \right]}$$

$$\phi_{i.1.u.hi}(f) := \text{atan}\left(\frac{\text{Im}\left(T_{i.1.u.hi}(f) \right)}{\text{Re}\left(T_{i.1.u.hi}(f) \right)} \right)$$

$$T_{o.1.u.hi}(f) := \frac{Z4(f)}{Z4(f) + Z3(f) + \left(\dfrac{1}{Z_{v2.hi}(f)} + \dfrac{1}{Z_{o.a.v1.u}(f)} \right)^{-1}}$$

$$\phi_{o.1.u.hi}(f) := \text{atan}\left(\frac{\text{Im}\left(T_{o.1.u.hi}(f) \right)}{\text{Re}\left(T_{o.1.u.hi}(f) \right)} \right)$$

$$T_{tot.83.u.hi}(f) := T_{i.1.u.hi}(f) T_{o.1.u.hi}(f) \cdot G_{aik.83.u.hi}$$

$$T_{tot.83.u.hi.e}(f) := 20 \cdot \log\left(\left| T_{tot.83.u.hi}(f) \right| \right)$$

$$T_{tot.83.u.hi.e}(h) = 39.621 \times 10^0 \quad [dB]$$

$$\phi_{tot.83.u.hi}(f) := \phi_{i.1.u.hi}(f) + \phi_{o.1.u.hi}(f) + \phi_{G.v1}(f)$$

9.10.14 Frequency and phase response plots:

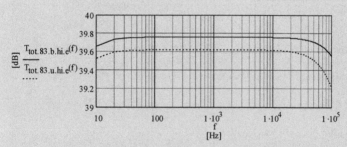

Figure 9.30 Frequency responses of the bypassed and un-bypassed
AIK gain stage

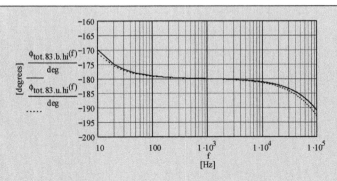

Figure 9.31 Phase responses of the bypassed and un-bypassed
AIK gain stage

9.10.15 Input impedances:

$$Z_{in.b.hi}(f) := \left[\left(R_{gg1.83} + \frac{1}{2j \cdot \pi \cdot f \cdot C_{i.v1.b.hi}} \right)^{-1} + \frac{1}{R_{g1.83}} \right]^{-1} + \frac{1}{2j \cdot \pi \cdot f \cdot C_{in.83}}$$

$$Z_{in.u.hi}(f) := \left[\left(R_{gg1.83} + \frac{1}{2j \cdot \pi \cdot f \cdot C_{i.v1.u.hi}} \right)^{-1} + \frac{1}{R_{g1.83}} \right]^{-1} + \frac{1}{2j \cdot \pi \cdot f \cdot C_{in.83}}$$

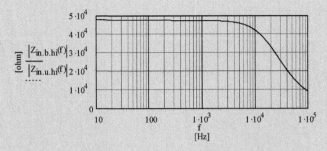

Figure 9.32 Input impedances for the bypassed and un-bypassed versions

Chapter 10 Cascoded Cathode Follower (CCF)

10.1 Circuit diagram

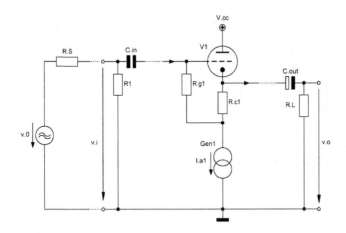

Figure 10.1 Basic design of the Cascoded Cathode Follower (CCF) gain stage

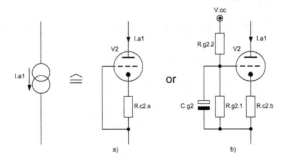

Figure 10.2 Constant current sink alternatives: a) = low-Z, b) = high-Z

10.2 Basic assumptions

The CCF is nothing else but a CF2$_u$ around V1 (improved CF version à la Chapter 3) with an actively loaded cathode by a constant current sink around V2 à la Chapters 5.2 or 5.4, thus, as mentioned in Chapter 3, producing a gain near 1 because of the very high internal resistance R_{v2} of the current sinks that replace the cathode resistor R_{c2} of Figure 3.4.

With this in mind and with the following assumptions the relevant equations become derivations of the equations given in the previous CF and CCG chapters:

- V1 does not need to equal V2
- $g_{m.v1}$ = or ≠ $g_{m.v2}$
- $r_{a.v1}$ = or ≠ $r_{a.v2}$
- μ_{v1} = or ≠ μ_{v2}
- R_{c1} = or ≠ $R_{c2.a}$
- $R_{c2.b}$ ≥ $R_{c2.a}$

But:
- plate current V1 I_{a1} = plate current V2 I_{a2}

10.3 Basic formulae (excl. stage load (R_L)-effect)[1]

10.3.1 Gain of the low-Z version:

The V2 active AC resistance $R_{v2.lo}$ becomes:

$$R_{v2.lo} = r_{a.v2} + \left(1 + \mu_{v2}\right)R_{c2.a} \tag{10.1}$$

The gain $G_{ccf.lo}$ of the low-Z loaded CCF$_{lo}$ becomes:

$$G_{ccf.lo} = g_{m.v1} \frac{r_{a.v1} \| \left(R_{c1} + R_{v2.lo}\right)}{1 + g_{m.v1}\left[r_{a.v1} \| \left(R_{c1} + R_{v2.lo}\right)\right]}$$

$$= \mu_{v1} \frac{R_{c1} + R_{v2.lo}}{r_{a.v1} + \left(1 + \mu_{v1}\right)\left(R_{c1} + R_{v2.lo}\right)} \tag{10.2}$$

rule of thumb:

with $R_{v2.lo} \gg r_{a.v1}$ and $R_{c2.a} \geq R_{c1}$ $G_{ccf.lo.rot}$ becomes:

$$G_{ccf.lo.rot} = \frac{\mu_{v1}}{1 + \mu_{v1}} \tag{10.3}$$

---

[1]   For detailed derivations of the shown equations: please refer to Chapter 3

The V2 active AC impedance $Z_{v2.lo}(f)$ becomes:

$$Z_{v2.lo}(f) = \left[\left(r_{a.v2} \parallel C_{a.c2} \right) + \left(1 + \mu_{v2} \right) \left(R_{c2.a} \parallel C_{g.c2} \right) \right] \parallel C_{g.a2} \parallel C_{stray.2} \quad (10.4)$$

10.3.2   Input and output impedances of the low-Z version:

Input resistance $R_{i.g1.lo}$ and impedance $Z_{i.g1.lo}(f)$ at the grid of V1 become:

$$R_{i.g1.lo} = \frac{R_{g1}}{1 - G_{ccf.lo}} \quad (10.5)$$

$$Z_{i.g1.lo}(f) = R_{i.g1.lo} \parallel C_{i.tot.lo} \quad (10.6)$$

$$C_{i.tot.lo} = \frac{\left[C_{g.a1} \parallel C_{stray.1} \parallel \left(1 - G_{ccf.lo} \right) C_{g.c1} \right] \left[C_{g.a2} \parallel C_{stray.2} \right]}{\left[C_{g.a1} \parallel C_{stray.1} \parallel \left(1 - G_{ccf.lo} \right) C_{g.c1} \right] + \left[C_{g.a2} \parallel C_{stray.2} \right]} \quad (10.7)$$

The output resistance $R_{o.c1.lo}$ at the cathode of V1 becomes[2]:

$$R_{o.c1.lo} = r_{c.v1} \parallel \left(R_{c1} + R_{v2.lo} \right) \quad (10.8)$$

$$r_{c.v1} = \frac{r_{a.v1}}{1 + \mu_{v1}} \quad (10.9)$$

10.3.3   Gain of the high-Z version:

The V2 active AC resistance $R_{v2.hi}$ becomes:

$$R_{v2.hi} = r_{a.v2} + \left(1 + \mu_{v2} \right) R_{c2.b} \quad (10.10)$$

The gain $G_{ccf.hi}$ of the low-Z loaded CCF_{hi} becomes:

$$\begin{aligned}
G_{ccf.hi} &= g_{m.v1} \frac{r_{a.v1} \parallel \left(R_{c1} + R_{v2.hi} \right)}{1 + g_{m.v1} \left[r_{a.v1} \parallel \left(R_{c1} + R_{v2.hi} \right) \right]} \\
&= \mu_{v1} \frac{R_{c1} + R_{v2.hi}}{r_{a.v1} + \left(1 + \mu_{v1} \right) \left(R_{c1} + R_{v2.hi} \right)}
\end{aligned} \quad (10.11)$$

rule of thumb:

with $R_{v2.hi} \gg r_{a.v1}$ and $R_{c2.b} \geq R_{c1}$ $G_{ccf.hi.rot}$ becomes:

---

[2]   Because of the rather low o/p resistance of V1 any valve related capacitance won't play a role in B_{20k}

$$G_{ccf.hi.rot} = \frac{\mu_{v1}}{1+\mu_{v1}} \tag{10.12}$$

The V2 active AC impedance $Z_{v2.lo}(f)$ becomes:

$$Z_{v2.hi}(f) = \left[\left(r_{a.v2} \parallel C_{a.c2} \right) + \left(1 + \mu_{v2} \right) \left(R_{c2.b} \parallel C_{g.c2} \right) \right] \parallel C_{g.a2} \parallel C_{stray.2} \tag{10.13}$$

10.3.4   Input and output impedances of the high-Z version:

Input resistance $R_{i.g1.hi}$ and impedance $Z_{i.g1.hi}(f)$ at the grid of V1 become:

$$R_{i.g1.hi} = \frac{R_{g1}}{1 - G_{ccf.hi}} \tag{10.14}$$

$$Z_{i.g1.hi}(f) = R_{i.g1.hi} \parallel C_{i.tot.hi} \tag{10.15}$$

$$C_{i.tot.hi} = \frac{\left[C_{g.a1} \parallel C_{stray.1} \parallel (1 - Ghi) C_{g.c1} \right]\left[C_{g.a2} \parallel C_{stray.2} \right]}{\left[C_{g.a1} \parallel C_{stray.1} \parallel (1 - G_{ccf.hi}) C_{g.c1} \right] + \left[C_{g.a2} \parallel C_{stray.2} \right]} \tag{10.16}$$

The output resistance $R_{o.c1.hi}$ at the cathode of V1 becomes[3]:

$$R_{o.c1.hi} = r_{c.v1} \parallel \left(R_{c1} + R_{v2.hi} \right) \tag{10.17}$$

$$r_{c.v1} = \frac{r_{a.v1}}{1+\mu_{v1}} \tag{10.18}$$

10.3.5   Grid voltage divider of the high-Z version and grid capacitance C_{g2}:

$$\begin{aligned} V_{C.g2} &= V_{R.c2.b} + V_{g2} \\ V_{R.c2.b} &= I_{a2} R_{c2.b} \end{aligned} \tag{10.19}$$

$$R_{g2.2} = R_{g2.1} \frac{V_{cc} - V_{C.g2}}{V_{C.g2}} \tag{10.20}$$

$$C_{g2} = \frac{1}{2\pi f_{c.opt} \left(R_{g2.1} \parallel R_{g2.2} \right)} \tag{10.21}$$

$$f_{c.opt} = 0.2 Hz \tag{10.22}$$

---

[3]   Because of the rather low o/p resistance of V1 any valve related capacitance won't play a role in B_{20k}

10.4 Gain stage frequency and phase response calculations

10.4.1 Low-Z version[4]:

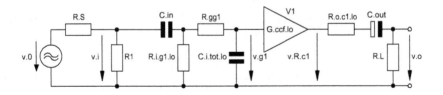

Figure 10.3 Simplified equivalent circuit of Figure 10.1 with the low-Z CCG as the cathode load - including all frequency and phase response relevant components

10.4.1.1   Gain stage input transfer function $T_{i.lo}(f)$ and phase $\varphi_{i.lo}(f)$ - including source resistance R_S and an oscillation preventing resistor $R_{gg1} \ll R_{g1}$:

$$T_{i.lo}(f) = \frac{v_{g1}}{v_0} \tag{10.23}$$

$$M_{i.lo}(f) = \frac{Z2_{lo}(f)}{R_{gg1} + Z2_{lo}(f)} \tag{10.24}$$

$$N_{i.lo}(f) = \frac{R_{i.g1.lo} \parallel \left(R_{gg1} + Z2_{lo}(f)\right)}{Z1(f) + \left[R_{i.g1.lo} \parallel \left(R_{gg1} + Z2_{lo}(f)\right)\right]} \tag{10.25}$$

$$O_{i.lo}(f) = \frac{R1 \parallel \left\{Z1(f) + \left[R_{i.g1.lo} \parallel \left(R_{gg1} + Z2_{lo}(f)\right)\right]\right\}}{RS + \left[R1 \parallel \left\{Z1(f) + \left[R_{i.g1.lo} \parallel \left(R_{gg1} + Z2_{lo}(f)\right)\right]\right\}\right]} \tag{10.26}$$

$$Z1(f) = \left(2j\pi f\, C_{in}\right)^{-1}$$
$$Z2_{lo}(f) = \left(2j\pi f\, C_{i.tot.lo}\right)^{-1} \tag{10.27}$$

$$T_{i.lo}(f) = M_{i.lo}(f)\, N_{i.lo}(f)\, O_{i.lo}(f) \tag{10.28}$$

$$\varphi_{i.lo}(f) = \arctan\left\{\frac{Im\left[T_{i.lo}(f)\right]}{Re\left[T_{i.lo}(f)\right]}\right\} \tag{10.29}$$

---

[4]   As could be demonstrated in the following MCD worksheets to achieve a gain near 1 this type of CCF works sufficiently well with high μ (>50) V1 valves. Because of their rather high r_a these valves also produce comparatively high cathode output resistances.

10.4.1.2   Gain stage output transfer function $T_{o.lo}(f)$ and phase $\varphi_{o.lo}(f)$:

$$Z3(f)=\left(2j\pi f\, C_{out}\right)^{-1} \tag{10.30}$$

$$T_{o.lo}(f)=\frac{v_o}{v_{R.cl}}=\frac{R_L}{R_L+R_{o.cl.lo}+Z3(f)} \tag{10.31}$$

$$\varphi_{o.lo}(f)=\arctan\left\{\frac{Im\left[T_{o.lo}(f)\right]}{Re\left[T_{o.lo}(f)\right]}\right\} \tag{10.32}$$

10.4.1.3   Fundamental gain stage phase shift $\varphi_{G.ccf.lo}(f)$:

$$\varphi_{G.ccf.lo}(f)=0° \tag{10.33}$$

10.4.1.4   Gain stage transfer function $T_{tot.lo}(f)$ and phase $\varphi_{tot.lo}(f)$:

$$T_{tot.lo}(f)=T_{i.lo}(f)\,T_{o.lo}(f)\,G_{ccf.lo} \tag{10.34}$$

$$\varphi_{tot.lo}(f)=\varphi_{i.lo}(f)+\varphi_{o.lo}(f)+\varphi_{G.ccf.lo}(f) \tag{10.35}$$

10.4.2   High-Z version[5]:

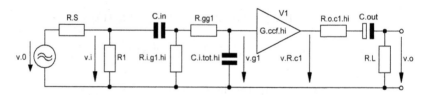

Figure 10.4 Simplified equivalent circuit of Figure 10.1 with the high-Z CCG as the cathode load - including all frequency and phase response relevant components

10.4.2.1   Gain stage input transfer function $T_{i.hi}(f)$ and phase $\varphi_{i.hi}(f)$ - including source resistance R_S and an oscillation preventing resistor $R_{gg1}\!\ll\!R_{g1}$:

$$T_{i.hi}(f)=\frac{v_{g1}}{v_0} \tag{10.36}$$

---

[5]   As could be demonstrated in the following MCD worksheets to achieve a gain near 1 this type of CCF works sufficiently well with low μ (<40) V1 valves. Because of their rather low r_a these valves also produce comparatively low cathode output resistances.

$$M_{i.hi}(f) = \frac{Z2_{hi}(f)}{R_{gg1} + Z2_{hi}(f)} \tag{10.37}$$

$$N_{i.hi}(f) = \frac{R_{i.g1.hi} \parallel \left(R_{gg1} + Z2_{hi}(f)\right)}{Z1(f) + \left[R_{i.g1.hi} \parallel \left(R_{gg1} + Z2_{hi}(f)\right)\right]} \tag{10.38}$$

$$O_{i.hi}(f) = \frac{R1 \parallel \left\{Z1(f) + \left[R_{i.g1.hi} \parallel \left(R_{gg1} + Z2_{hi}(f)\right)\right]\right\}}{RS + \left[R1 \parallel \left\{Z1(f) + \left[R_{i.g1.hi} \parallel \left(R_{gg1} + Z2_{hi}(f)\right)\right]\right\}\right]} \tag{10.39}$$

$$Z1(f) = \left(2j\pi f\, C_{in}\right)^{-1}$$
$$Z2_{hi}(f) = \left(2j\pi f\, C_{i.tot.hi}\right)^{-1} \tag{10.40}$$

$$T_{i.hi}(f) = M_{i.hi}(f)\, N_{i.hi}(f)\, O_{i.hi}(f) \tag{10.41}$$

$$\varphi_{i.hi}(f) = \arctan\left\{\frac{\mathrm{Im}\left[T_{i.hi}(f)\right]}{\mathrm{Re}\left[T_{i.hi}(f)\right]}\right\} \tag{10.42}$$

10.4.2.2   Gain stage output transfer function $T_{o.hi}(f)$ and phase $\varphi_{o.hi}(f)$:

$$Z3(f) = \left(2j\pi f\, C_{out}\right)^{-1} \tag{10.43}$$

$$T_{o.hi}(f) = \frac{v_o}{v_{R.cl}} = \frac{R_L}{R_L + R_{o.cl.hi} + Z3(f)} \tag{10.44}$$

$$\varphi_{o.hi}(f) = \arctan\left\{\frac{\mathrm{Im}\left[T_{o.hi}(f)\right]}{\mathrm{Re}\left[T_{o.hi}(f)\right]}\right\} \tag{10.45}$$

10.4.2.3   Fundamental gain stage phase shift $\varphi_{G.ccf.hi}(f)$:

$$\varphi_{G.ccf.hi}(f) = 0° \tag{10.46}$$

10.4.4.4   Gain stage transfer function $T_{tot.hi}(f)$ and phase $\varphi_{tot.hi}(f)$:

$$T_{tot.hi}(f) = T_{i.hi}(f)\, T_{o.hi}(f)\, G_{ccf.hi} \tag{10.47}$$

$$\varphi_{tot.hi}(f) = \varphi_{i.hi}(f) + \varphi_{o.hi}(f) + \varphi_{G.ccf.hi}(f) \tag{10.48}$$

10.5   Low-Z CCF example with V1,2 = ECC83 / 12AX7:

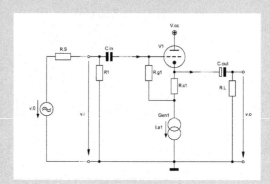

Figure 10.5 CCF-lo example circuitry

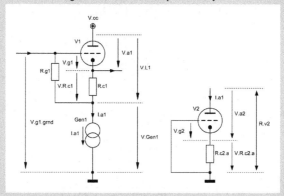

Figure 10.6 DC voltage arrangement of Figure 10.5 - lo version

10.5.1   Triodes bias data:

$$I_{a1} := 1.2 \cdot 10^{-3} A \qquad I_{a2} := I_{a1}$$

$$V_{cc} := 350 V \qquad V_{Gen1} := 100.5 V \qquad V_{g1} := -2.0 V \qquad V_{g2} := -0.5 V$$

$$V_{L1} := V_{cc} - V_{Gen1} \qquad V_{L1} = 249.5 V \qquad V_{R.c1} := |V_{g1}| \qquad V_{R.c2.a} := |V_{g2}|$$

$$V_{a1} := V_{L1} - V_{R.c1} \qquad V_{a1} = 247.5 V \qquad V_{a2} := V_{Gen1} - V_{R.c2.a} \qquad V_{a2} = 100 V$$

$$V_{g1.grnd} := V_{Gen1} \qquad V_{g1.grnd} = 100.5 V$$

10.5.2. Triodes valve constants:

$$g_{m.v1} := 1.6 \cdot 10^{-3} \cdot S \qquad r_{a.v1} := 62.5 \cdot 10^3 \Omega \qquad \mu_{v1} := 100$$

$$g_{m.v2} := 1.75 \cdot 10^{-3} S \qquad r_{a.v2} := 58 \cdot 10^3 \Omega \qquad \mu_{v2} := 101.5$$

$$C_{g.c1.83} := 1.65 \cdot 10^{-12} F \qquad C_{g.a1.83} := 1.6 \cdot 10^{-12} F \qquad C_{a.c1.83} := 0.33 \cdot 10^{-12} F$$

$$C_{g.c2.83} := 1.65 \cdot 10^{-12} F \qquad C_{g.a2.83} := 1.6 \cdot 10^{-12} F \qquad C_{a.c2.83} := 0.33 \cdot 10^{-12} F$$

10.5.3 Circuit variables:

$$C_{stray.1} := 10 \cdot 10^{-12} F \qquad C_{stray.2} := 10 \cdot 10^{-12} F$$

$$R_{c1} := \frac{V_{R.c1}}{I_{a1}} \qquad R_{c1} = 1.667 \times 10^3 \Omega \qquad R_{gg1} := 2.21 \cdot 10^3 \Omega$$

$$R_{c2.a} := \frac{V_{R.c2.a}}{I_{a2}} \qquad R_{c2.a} = 416.667 \Omega$$

$$RS := 1 \cdot 10^3 \Omega \qquad R_{g1} := 1 \cdot 10^6 \Omega \qquad R_L := 100 \cdot 10^3 \Omega$$

$$R1 := 100 \cdot 10^3 \Omega$$

$$C_{in} := 0.1 \cdot 10^{-6} F \qquad C_{out} := 100 \cdot 10^{-6} F$$

10.5.4 Calculation relevant data:

frequency range f for the below shown graphs: $\qquad f := 10Hz, 20Hz .. 100000\,Hz$

$$h := 1000Hz$$

10.5.5 Gain $G_{ccf.lo}$ (frequency independent):

$$R_{v2.lo} := r_{a.v2} + (1 + \mu_{v2}) \cdot R_{c2.a} \qquad\qquad R_{v2.lo} = 1.007 \times 10^5 \Omega$$

$$G_{ccf.lo.83.1} := g_{m.v1} \cdot \frac{\left(\dfrac{1}{r_{a.v1}} + \dfrac{1}{R_{c1} + R_{v2.lo}} \right)^{-1}}{1 + g_{m.v1} \cdot \left(\dfrac{1}{r_{a.v1}} + \dfrac{1}{R_{c1} + R_{v2.lo}} \right)^{-1}} \qquad\qquad G_{ccf.lo.83.1} = 0.984$$

➤   MCD Worksheet X-1   CCF-lo calculations   Page 3

$$G_{ccf.lo.83.2} := \mu_{v1} \cdot \frac{R_{c1} + R_{v2.lo}}{r_{a.v1} + (1 + \mu_{v1}) \cdot (R_{c1} + R_{v2.lo})}$$

$$G_{ccf.lo.83.2} = 0.984$$

$$G_{ccf.lo.83.1.e} := 20 \cdot \log(G_{ccf.lo.83.1})$$

$$G_{ccf.lo.83.1.e} = -0.139 \quad [dB]$$

$$G_{ccf.lo.83.rot} := \frac{\mu_{v1}}{1 + \mu_{v1}}$$

$$G_{ccf.lo.83.rot} = 0.99$$

$$G_{ccf.lo.83.rot.e} := 20 \cdot \log(G_{ccf.lo.83.rot})$$

$$G_{ccf.lo.83.rot.e} = -0.086 \quad [dB]$$

$$G_{ccf.lo.83.1.e} - G_{ccf.lo.83.rot.e} = -0.052 \quad [dB]$$

10.5.6 Gain $G_{ccf.lo}(f)$ (frequency dependent):

$$X_{v2.lo}(f) := 2j \cdot \pi \cdot f \cdot (C_{g.a2.83} + C_{stray.2})$$

$$Y_{v2.lo}(f) := \left[\left(\frac{1}{r_{a.v2}} + 2j \cdot \pi \cdot f \cdot C_{a.c2.83} \right)^{-1} + (1 + \mu_{v2}) \cdot \left(\frac{1}{R_{c2.a}} + 2j \cdot \pi \cdot f \cdot C_{g.c2.83} \right)^{-1} \right]^{-1}$$

$$Z_{v2.lo}(f) := \left[(Y_{v2.lo}(f)) + X_{v2.lo}(f) \right]^{-1}$$

$$|Z_{v2.lo}(h)| = 1.007 \times 10^5 \, \Omega$$

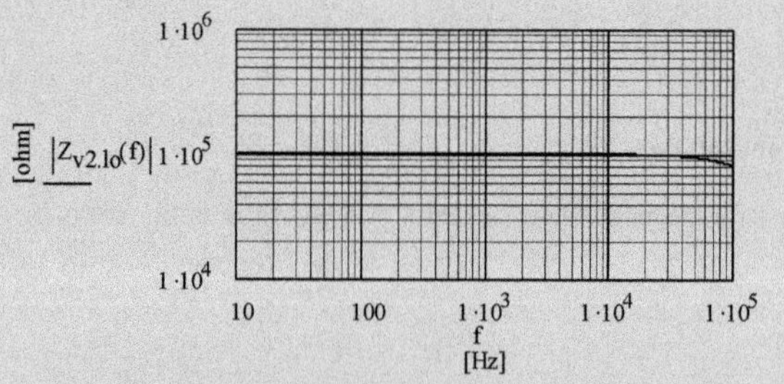

Figure 10.7 Impedance of the low-Z constant current sink

$$G_{ccf.lo.83}(f) := \mu_{v1} \cdot \frac{R_{c1} + Z_{v2.lo}(f)}{r_{a.v1} + (1 + \mu_{v1}) \cdot (R_{c1} + Z_{v2.lo}(f))}$$

$$|G_{ccf.lo.83}(h)| = 0.984$$

$$G_{ccf.lo.83.e}(f) := 20 \cdot \log(|G_{ccf.lo.83}(f)|)$$

$$G_{ccf.lo.83.e}(h) = -0.139 \quad [dB]$$

$$\phi_{ccf.lo.83}(f) := \frac{Im(G_{ccf.lo.83}(f))}{Re(G_{ccf.lo.83}(f))}$$

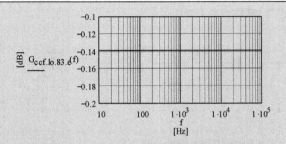

Figure 10.8 CCF-lo gain vs. frequency

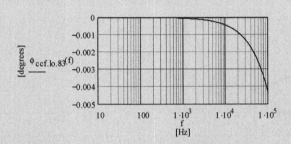

Figure 10.9 CCF-lo phase vs. frequency

10.5.7 Specific resistances and capacitances:

$$C_{i.tot.lo} := \left[\frac{1}{\left[\left(1 - G_{ccf.lo.83.1}\right) \cdot C_{g.c1.83} + C_{stray.1} + C_{g.a1.83} \right]} + \frac{1}{\left(C_{g.a2.83} + C_{stray.2} \right)} \right]^{-1}$$

$$C_{i.tot.lo} = 5.807 \times 10^{-12}\,F$$

$$R_{i.g1.lo} := \frac{R_{g1}}{1 - G_{ccf.lo.83.1}} \qquad\qquad R_{i.g1.lo} = 6.309 \times 10^{7}\,\Omega$$

$$r_{c.v1} := \frac{r_{a.v1}}{1 + \mu_{v1}} \qquad\qquad r_{c.v1} = 618.812\,\Omega$$

$$R_{o.c1.lo} := \left(\frac{1}{r_{c.v1}} + \frac{1}{R_{c1} + R_{v2.lo}} \right)^{-1} \qquad\qquad R_{o.c1.lo} = 615.094\,\Omega$$

10.5.8 Gain stage frequency and phase response:

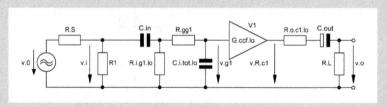

Figure 10.10 = Figure 10.3

$$Z1(f) := \left(2j \cdot \pi \cdot f \cdot C_{in}\right)^{-1} \qquad\qquad Z2_{lo}(f) := \left(2j \cdot \pi \cdot f \cdot C_{i.tot.lo}\right)^{-1}$$

$$M_{i.lo}(f) := \frac{Z2_{lo}(f)}{R_{gg1} + Z2_{lo}(f)}$$

$$N_{i.lo}(f) := \frac{\left[\left(R_{i.g1.lo}\right)^{-1} + \left(R_{gg1} + Z2_{lo}(f)\right)^{-1}\right]^{-1}}{Z1(f) + \left[\left(R_{i.g1.lo}\right)^{-1} + \left(R_{gg1} + Z2_{lo}(f)\right)^{-1}\right]^{-1}}$$

$$O_{i.lo}(f) := \frac{\left[\dfrac{1}{R1} + \dfrac{1}{Z1(f) + \left[\left(R_{i.g1.lo}\right)^{-1} + \left(R_{gg1} + Z2_{lo}(f)\right)^{-1}\right]^{-1}}\right]^{-1}}{RS + \left[\dfrac{1}{R1} + \dfrac{1}{Z1(f) + \left[\left(R_{i.g1.lo}\right)^{-1} + \left(R_{gg1} + Z2_{lo}(f)\right)^{-1}\right]^{-1}}\right]^{-1}}$$

$$T_{i.lo}(f) := M_{i.lo}(f) \cdot N_{i.lo}(f) \cdot O_{i.lo}(f) \qquad\qquad \phi_{i.lo}(f) := atan\left(\frac{Im\left(T_{i.lo}(f)\right)}{Re\left(T_{i.lo}(f)\right)}\right)$$

$$T_{i.lo.e}(f) := 20 \cdot \log\left(\left|T_{i.lo}(f)\right|\right)$$

$$Z3(f) := \left(2j \cdot \pi \cdot f \cdot C_{out}\right)^{-1}$$

$$T_{o.lo}(f) := \frac{R_L}{R_L + R_{o.c1.lo} + Z3(f)} \qquad\qquad \phi_{o.lo}(f) := atan\left(\frac{Im\left(T_{o.lo}(f)\right)}{Re\left(T_{o.lo}(f)\right)}\right)$$

$$T_{o.lo.e}(f) := 20 \cdot \log\left(\left|T_{o.lo}(f)\right|\right)$$

➤   MCD Worksheet X-1   CCF-lo calculations   Page 6

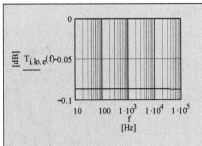

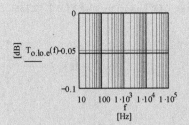

Figure 10.11 Transfer of lo i/p network   Figure 10.12 Transfer of lo o/p network

$$T_{tot.lo}(f) := T_{i.lo}(f) \cdot T_{o.lo}(f) \cdot G_{ccf.lo.83.1}$$ $$\phi_{G.ccf.lo}(f) := 0$$

$$T_{tot.lo.e}(f) := 20 \cdot \log\left(\left|T_{tot.lo}(f)\right|\right)$$

$$\phi_{tot.lo}(f) := \phi_{i.lo}(f) + \phi_{o.lo}(f) + \phi_{G.ccf.lo}(f)$$

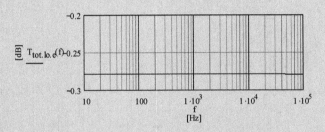

Figure 10.13 Frequency response of the CCF-lo gain stage

$$T_{tot.lo.e}(h) := 20 \cdot \log\left(\left|T_{tot.lo}(h)\right|\right)$$ $$T_{tot.lo.e}(h) = -0.279 \ [dB]$$

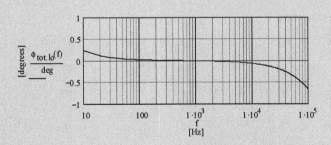

Figure 10.14 Phase response of the CCF-lo gain stage

➢ **MCD Worksheet X-2   CCF-hi calculations   Page 1**

10.6   High-Z CCF example with V1, V2 = E188CC / 7308:

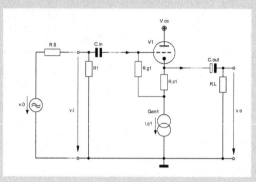

Figure 10.15 CCF-hi example circuitry

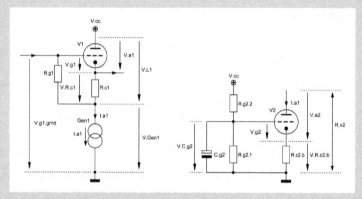

Figure 10.16 DC voltage arrangement of Figure 10.15 - hi version

10.6.1 Triodes bias data:

$I_{a1} := 10 \cdot 10^{-3} A$ $\qquad$ $I_{a2} := I_{a1}$ $\qquad$ $V_{a1} := 90V$ $\qquad$ $V_{a2} := 90V$

$V_{cc} := 350V$ $\qquad\qquad\qquad$ $V_{g1} := -1.65V$ $\qquad$ $V_{g2} := -1.65V$

$V_{R.c1} := |V_{g1}|$ $\qquad$ $V_{Gen1} := V_{cc} - (V_{a1} + V_{R.c1})$ $\qquad$ $V_{Gen1} = 258.35V$

$V_{R.c2.b} := V_{Gen1} - V_{a2}$ $\quad$ $V_{R.c2.b} = 168.35V$ $\quad$ $R_{c2.b} := \dfrac{V_{R.c2.b}}{I_{a2}}$ $\quad$ $R_{c2.b} = 1.684 \times 10^{4}\Omega$

$V_{g1.grnd} := V_{Gen1}$ $\quad$ $V_{g1.grnd} = 258.35V$ $\quad$ $V_{C.g2} := V_{R.c2.b} + V_{g2}$ $\quad$ $V_{C.g2} = 166.7V$

$R_{g2.1} := 1 \cdot 10^{6}\Omega$ $\qquad$ $R_{g2.2} := R_{g2.1} \dfrac{(V_{cc} - V_{C.g2})}{V_{C.g2}}$ $\qquad\qquad$ $R_{g2.2} = 1.1 \times 10^{6}\Omega$

➢ MCD Worksheet X-2   CCF-hi calculations   Page 2

10.6.2. Triodes valve constants:

$g_{m.v1} := 10.4 \cdot 10^{-3} \cdot S$ $r_{a.v1} := 3.1 \cdot 10^3 \Omega$ $\mu_{v1} := 32$

$g_{m.v2} := 10.4 \cdot 10^{-3} S$ $r_{a.v2} := 3.1 \cdot 10^3 \Omega$ $\mu_{v2} := 32$

$C_{g.c1.188} := 3.1 \cdot 10^{-12} F$ $C_{g.a1.188} := 1.4 \cdot 10^{-12} F$ $C_{a.c1.188} := 1.75 \cdot 10^{-12} F$

$C_{g.c2.188} := 3.1 \cdot 10^{-12} F$ $C_{g.a2.188} := 1.4 \cdot 10^{-12} F$ $C_{a.c2.188} := 1.75 \cdot 10^{-12} F$

10.6.3 Circuit variables:

$C_{stray.1} := 10 \cdot 10^{-12} F$ $C_{stray.2} := 10 \cdot 10^{-12} F$

$R_{c1} := \dfrac{V_{R.c1}}{I_{a1}}$ $R_{c1} = 165 \, \Omega$ $R_{gg1} := 2.21 \cdot 10^3 \Omega$

$RS := 1 \cdot 10^3 \Omega$ $R_{g1} := 1 \cdot 10^6 \Omega$ $R_L := 100 \cdot 10^3 \Omega$

$R1 := 100 \cdot 10^3 \Omega$

$C_{in} := 0.1 \cdot 10^{-6} F$ $C_{out} := 100 \cdot 10^{-6} F$ $f_{c.opt} := 0.2 Hz$

$C_{g2} := \dfrac{1}{2 \cdot \pi \cdot f_{c.opt} \cdot \left(\dfrac{1}{R_{g2.1}} + \dfrac{1}{R_{g2.2}} \right)^{-1}}$ $C_{g2} = 1.519 \times 10^{-6} F$

10.6.4 Calculation relevant data:

frequency range f for the below shown graphs: $f := 10Hz, 20Hz .. 100000\,Hz$

$h := 1000Hz$

10.6.5 Gain $G_{ccf.hi}$ (frequency independent):

$R_{v2.hi} := r_{a.v2} + \left(1 + \mu_{v2} \right) \cdot R_{c2.b}$ $R_{v2.hi} = 5.587 \times 10^5 \, \Omega$

$G_{ccf.hi.188.1} := g_{m.v1} \cdot \dfrac{\left(\dfrac{1}{r_{a.v1}} + \dfrac{1}{R_{c1} + R_{v2.hi}} \right)^{-1}}{1 + g_{m.v1} \cdot \left(\dfrac{1}{r_{a.v1}} + \dfrac{1}{R_{c1} + R_{v2.hi}} \right)^{-1}}$ $G_{ccf.hi.188.1} = 0.97$

➢ MCD Worksheet X-2 CCF-hi calculations Page 3

$$G_{ccf.hi.188.2} := \mu_{v1} \cdot \frac{R_{c1} + R_{v2.hi}}{r_{a.v1} + (1 + \mu_{v1}) \cdot (R_{c1} + R_{v2.hi})}$$ $G_{ccf.hi.188.2} = 0.97$

$$G_{ccf.hi.188.1.e} := 20 \cdot \log(G_{ccf.hi.188.1})$$ $G_{ccf.hi.188.1.e} = -0.267$ [dB]

$$G_{ccf.hi.188.rot} := \frac{\mu_{v1}}{1 + \mu_{v1}}$$ $G_{ccf.hi.188.rot} = 0.97$

$$G_{ccf.hi.188.rot.e} := 20 \cdot \log(G_{ccf.hi.188.rot})$$ $G_{ccf.hi.188.rot.e} = -0.267$ [dB]

$$G_{ccf.hi.188.1.e} - G_{ccf.hi.188.rot.e} = 5.101 \times 10^{-4} \quad \text{[dB]}$$

10.6.6 Gain $G_{ccf.hi}(f)$ (frequency dependent):

$$X_{v2.hi}(f) := 2j \cdot \pi \cdot f \cdot (C_{g.a2.188} + C_{stray.2})$$

$$Y_{v2.hi}(f) := \left[\left(\frac{1}{r_{a.v2}} + 2j \cdot \pi \cdot f \cdot C_{a.c2.188}\right)^{-1} + (1 + \mu_{v2}) \cdot \left(\frac{1}{R_{c2.b}} + 2j \cdot \pi \cdot f \cdot C_{g.c2.188}\right)^{-1}\right]^{-1}$$

$$Z_{v2.hi}(f) := \left[(Y_{v2.hi}(f)) + X_{v2.hi}(f)\right]^{-1}$$ $|Z_{v2.hi}(h)| = 5.582 \times 10^5 \Omega$

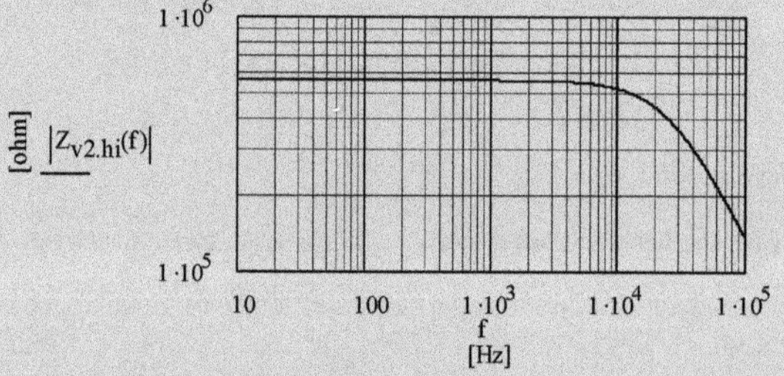

Figure 10.17 Impedance of the high-Z constant current sink

$$G_{ccf.hi.188}(f) := \mu_{v1} \cdot \frac{R_{c1} + Z_{v2.hi}(f)}{r_{a.v1} + (1 + \mu_{v1}) \cdot (R_{c1} + Z_{v2.hi}(f))}$$ $|G_{ccf.hi.188}(h)| = 0.97$

$$G_{ccf.hi.188.e}(f) := 20 \cdot \log(|G_{ccf.hi.188}(f)|)$$ $G_{ccf.hi.188.e}(h) = -0.269$ [dB]

$$\phi_{ccf.hi.188}(f) := \frac{\mathrm{Im}(G_{ccf.hi.188}(f))}{\mathrm{Re}(G_{ccf.hi.188}(f))}$$

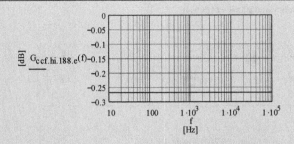

Figure 10.18 CCF-hi gain vs. frequency

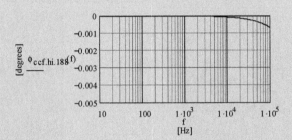

Figure 10.19 CCF-hi phase vs. frequency

10.6.7 Specific resistances and capacitances:

$$C_{i.tot.hi} := \left[\frac{1}{\left[\left(1 - G_{ccf.hi.188.1} \right) \cdot C_{g.c1.188} + C_{stray.1} + C_{g.a1.188} \right]} + \frac{1}{\left(C_{g.a2.188} + C_{stray.2} \right)} \right]^{-1}$$

$$C_{i.tot.hi} = 5.723 \times 10^{-12} F$$

$$R_{i.g1.hi} := \frac{R_{g1}}{1 - G_{ccf.hi.188.1}} \qquad\qquad R_{i.g1.hi} = 3.306 \times 10^{7} \, \Omega$$

$$r_{c.v1} := \frac{r_{a.v1}}{1 + \mu_{v1}} \qquad\qquad r_{c.v1} = 93.939 \, \Omega$$

$$R_{o.c1.hi} := \left(\frac{1}{r_{c.v1}} + \frac{1}{R_{c1} + R_{v2.hi}} \right)^{-1} \qquad\qquad R_{o.c1.hi} = 93.924 \, \Omega$$

10.6.8 Gain stage frequency and phase response:

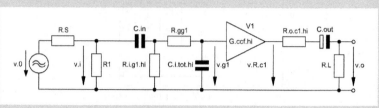

Figure 10.20 = Figure 10.4

$$Z1(f) := \left(2j \cdot \pi \cdot f \cdot C_{in}\right)^{-1} \qquad\qquad Z2_{hi}(f) := \left(2j \cdot \pi \cdot f \cdot C_{i.tot.hi}\right)^{-1}$$

$$M_{i.hi}(f) := \frac{Z2_{hi}(f)}{R_{gg1} + Z2_{hi}(f)}$$

$$N_{i.hi}(f) := \frac{\left[\left(R_{i.g1.hi}\right)^{-1} + \left(R_{gg1} + Z2_{hi}(f)\right)^{-1}\right]^{-1}}{Z1(f) + \left[\left(R_{i.g1.hi}\right)^{-1} + \left(R_{gg1} + Z2_{hi}(f)\right)^{-1}\right]^{-1}}$$

$$O_{i.hi}(f) := \frac{\left[\dfrac{1}{R1} + \dfrac{1}{Z1(f) + \left[\left(R_{i.g1.hi}\right)^{-1} + \left(R_{gg1} + Z2_{hi}(f)\right)^{-1}\right]^{-1}}\right]^{-1}}{RS + \left[\dfrac{1}{R1} + \dfrac{1}{Z1(f) + \left[\left(R_{i.g1.hi}\right)^{-1} + \left(R_{gg1} + Z2_{hi}(f)\right)^{-1}\right]^{-1}}\right]^{-1}}$$

$$T_{i.hi}(f) := M_{i.hi}(f) \cdot N_{i.hi}(f) \cdot O_{i.hi}(f) \qquad\qquad \phi_{i.hi}(f) := atan\left(\frac{Im\left(T_{i.hi}(f)\right)}{Re\left(T_{i.hi}(f)\right)}\right)$$

$$T_{i.hi.e}(f) := 20 \cdot log\left(\left|T_{i.hi}(f)\right|\right)$$

$$Z3(f) := \left(2j \cdot \pi \cdot f \cdot C_{out}\right)^{-1}$$

$$T_{o.hi}(f) := \frac{R_L}{R_L + R_{o.c1.hi} + Z3(f)} \qquad\qquad \phi_{o.hi}(f) := atan\left(\frac{Im\left(T_{o.hi}(f)\right)}{Re\left(T_{o.hi}(f)\right)}\right)$$

$$T_{o.hi.e}(f) := 20 \cdot log\left(\left|T_{o.hi}(f)\right|\right)$$

➢   MCD Worksheet X-2   CCF-hi calculations   Page 6

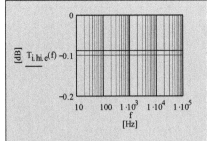

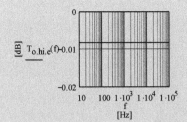

Figure 10.21 Transfer of hi i/p network   Figure 10.22 Phase of hi o/p network

$$T_{tot.hi}(f) := T_{i.hi}(f) \cdot T_{o.hi}(f) \cdot G_{ccf.hi} \cdot 188.1$$ $$\phi_{G.ccf.hi}(f) := 0$$

$$T_{tot.hi.e}(f) := 20 \cdot \log\left(\left|T_{tot.hi}(f)\right|\right)$$

$$\phi_{tot.hi}(f) := \phi_{i.hi}(f) + \phi_{o.hi}(f) + \phi_{G.ccf.hi}(f)$$

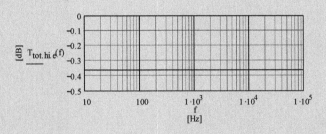

Figure 10.23 Frequency response of the CCF-hi gain stage

$$T_{tot.hi.e}(h) := 20 \cdot \log\left(\left|T_{tot.hi}(h)\right|\right)$$ $$T_{tot.hi.e}(h) = -0.362 \quad [dB]$$

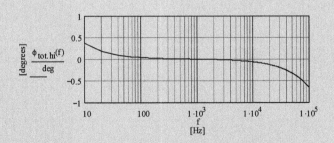

Figure 10.24 Phase response of the CCF-hi gain stage

Chapter 11 White Cathode Follower (WCF)

11.1 Circuit diagram

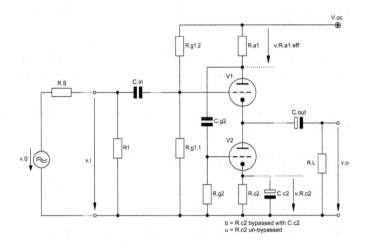

Figure 11.1 Basic design of the White Cathode Follower gain stage (WCF)

11.2 Remarks and basic assumptions

The WCF's main aim is to achieve a very much lower output resistance than that of the CF or CCF - by keeping its gain (alike the CCF) close to 1.

The WCF is a CCF à la Chapter 10 (V1) with a positive feedback path from V1's plate via V2 back into the cathode of V1, thus, creating a rather small output resistance at the cathode of V1 - very much smaller than the output resistance of an ordinary CCF without that feedback loop.

The V2 – V1 sequence looks like a CAS gain stage à la Chapter 6. Hence, the whole design can also be described as a CCF with an additional CAS gain stage as feedback loop. V2's grid plays the input of the CAS. Cutting-off the feedback loop (eg. $R_{a1} = 0R$ or no C_{g2}) will automatically lead to a CCF.

Because, in principle, the gain of the whole stage never exceeds the 0dB line the positive feedback should not be able to force the stage into oscillation.

Equal DC operation of V1 and V2 should ensure rather symmetrical signal output voltage into any load R_L. To avoid clipping maximal signal output voltage and current should be based on the current delivery possibilities of the output of the gain stage! In addition, at the grid of V2 any signal peak voltage that is bigger than the grid's DC voltage for proper operation of the valve will bring V2 into a cut-off state.

To enable easy calculations the following assumptions have to be taken into account:

- V1 equals V2 (ideal case: double triode)
- plate current V1   = plate current V2 $(I_{a1} = I_{a2})$
- plate-cathode voltage V1   = plate cathode voltage V2 $(V_{a1} = V_{a2})$
- g_m $= g_{m.v1}$ $= g_{m.v2}$
- r_a $= r_{a.v1}$ $= r_{a.v2}$
- μ $= \mu_{v1}$ $= \mu_{v2}$
- bypassed version (b): R_{c2} = bypassed by C_{c2}
- un-bypassed version (u): R_{c2} = un-bypassed

11.3 Basic formulae for both versions (excl. and incl. stage load (R_L)-effect)

$$R_{a1.eff} = R_{a1} \parallel R_{g2} \tag{11.1}$$

11.3.1 Gain $G_{wcf.b.nL}$ excl. output load (nL) (bypassed version):

$$G_{wcf.b.nL} = \frac{\mu\left(r_a + \mu R_{a1.eff}\right)}{r_a\left(\mu + 2\right) + R_{a1.eff}\left(\mu^2 + \mu + 1\right)} \tag{11.2}$$

11.3.2 Gain $G_{wcf.b.L}$ incl. output load (L) (bypassed version):

$$G_{wcf.b.L} = \frac{\mu\left(r_a + \mu R_{a1.eff}\right)}{\dfrac{r_a\left(r_a + \mu R_{a1.eff}\right)}{R_L} + r_a\left(\mu + 2\right) + R_{a1.eff}\left(\mu^2 + \mu + 1\right)} \tag{11.3}$$

11.3.3 Output resistance $R_{o.b}$ (bypassed version):

$$R_{o.b} = \frac{r_a\left(r_a + R_{a1.eff}\right)}{r_a\left(\mu + 2\right) + R_{a1.eff}\left(\mu^2 + \mu + 1\right)} \tag{11.4}$$

Rule of thumb[1]:

$$
\begin{aligned}
R_{o.b.rot} &\approx \frac{r_{c1}}{\left|G_{cas.b.rot}\right|} \\
&\approx \left(\frac{r_a}{\mu R_{a1}}\right)\left(\frac{r_a + R_{a1}}{\mu + 1}\right)
\end{aligned}
\tag{11.5}
$$

11.3.4 Gain $G_{wcf.u.nL}$ excl. output load (nL) (un-bypassed version):

$$G_{wcf.u.nL} = \frac{\mu r_a + \mu^2 R_{a1.eff} + \left(\mu^2 + \mu\right)R_{c2}}{\left(\mu + 2\right)r_a + \left(\mu^2 + \mu + 1\right)R_{a1.eff} + \left(\mu^2 + 2\mu + 1\right)R_{c2}} \tag{11.6}$$

11.3.5 Gain $G_{wcf.u.L}$ incl. output load (L) (un-bypassed version):

$$G_{wcf.u.L} = \frac{\mu r_a + \mu^2 R_{a1.eff} + \left(\mu^2 + \mu\right)R_{c2}}{\left(\dfrac{A}{R_L} + B\right)} \tag{11.7}$$

---

[1] to get $G_{cas.b.rot}$ see Chapter 6

$$A = r_a \left[r_a + R_{a1.eff} + (\mu + 1) R_{c2} \right] + (\mu + 1) R_{c2} R_{a1.eff}$$

$$B = (\mu + 2) r_a + \left(\mu^2 + \mu + 1 \right) R_{a1.eff} + \left(\mu^2 + 2\mu + 1 \right) R_{c2}$$

(11.8)

11.3.6   Output resistance $R_{o.u}$ (un-bypassed version):

$$R_{o.u} = \frac{\left(r_a + R_{a1.eff} \right) \left[r_a + R_{c2} (\mu + 1) \right]}{(\mu + 2) r_a + \left(\mu^2 + \mu + 1 \right) R_{a1.eff} + \left(\mu^2 + 2\mu + 1 \right) R_{c2}}$$

(11.9)

Rule of thumb: n.a.

11.3.7   Specific resistances and capacitances[2]:

11.3.7.1   Both versions -   input resistance R_i & impedance $Z_i(f)$, input & output capacitance C_{in} & C_{out}, grid 2 capacitance C_{g2}:

$$R_i = R1 \parallel R_{g1.1} \parallel R_{g1.2}$$

(11.10)

$$Z_i(f) = R_i \parallel C_i$$

(11.11)

$$C_{in} = \frac{1}{2\pi f_{c.opt} \left(R_{g1} \parallel R_{g2} \right)}$$

(11.12)

$$C_{out} = \frac{1}{2\pi f_{c.opt} R_L}$$

(11.13)

$$C_{g2} = \frac{1}{2\pi f_{c.opt} R_{g2}}$$

(11.14)

$$f_{c.opt} = 0.2 Hz$$

(11.15)

11.3.7.2   version (b) only - cathode capacitance C_{c2}:

$$C_{c2} = \frac{1}{2\pi f_{c.opt} \left(r_{c2} \parallel R_{c2} \right)}$$

(11.16)

$$r_{c2} = \frac{R_{a1.eff} + 2 r_a}{\mu + 1}$$

(11.17)

---

[2]   For C_i see previous Chapters

11.4 Derivations

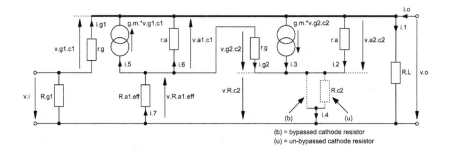

Figure 11.2 Equivalent circuit of Figure 11.1

11.4.1 Bypassed version:

11.4.1.1   Gains $G_{wcf.b}$ (output loaded with R_L or non-loaded):

$$G_{wcf.b} = \frac{v_o}{v_i} \qquad (11.18)$$

sum of currents at the top node:

$$i_o = i_1 + i_2 + i_3 - i_5 - i_6$$
$$\text{set } i_o = 0 \qquad (11.19)$$

$$\Rightarrow i_1 = \frac{v_o}{R_L} = i_5 + i_6 - i_2 - i_3$$

hence,

$$\frac{v_o}{R_L} = g_m v_{g1.c1} + \frac{v_{a1.c1}}{r_a} - g_m v_{g2.c2} - \frac{v_{a2.c2}}{r_a}$$

$$= \left(\begin{array}{l} g_m \left(v_i - v_o\right) - \dfrac{2v_o + v_{R.a1.eff}}{r_a} \\[2ex] + g_m R_{a1.eff} \left(g_m \left(v_i - v\right)_o - \dfrac{v_o + v_{R.a1.eff}}{r_a} \right) \end{array} \right) \qquad (11.20)$$

with:

$$v_{a2.c2} = v_o$$
$$v_{g2.c2} = -v_{R.al.eff}$$
$$v_{g1.c1} = v_i - v_o \tag{11.21}$$
$$v_{al.c1} = -v_o - v_{R.al.eff}$$

and:

$$
\begin{aligned}
v_{R.al.eff} &= i_7\, R_{al.eff} \\
&= (i_5 + i_6)\, R_{al.eff} \\
&= R_{al.eff} \left(g_m v_{g1.c1} + \frac{v_{al.c1}}{r_a} \right)
\end{aligned} \tag{11.22}
$$

we'll get:

$$v_{R.al.eff} = R_{al.eff}\, \frac{\mu\, v_i - (\mu + 1)\, v_o}{r_a + R_{al.eff}} \tag{11.23}$$

and - after a lot of rearrangements - we'll get the gain $G_{wcf.b.L}$ that includes the output load R_L:

$$G_{wcf.b.L} = \frac{\mu\left(r_a + \mu R_{al.eff}\right)}{\dfrac{r_a\left(r_a + R_{al.eff}\right)}{R_L} + r_a\left(\mu + 2\right) + R_{al.eff}\left(\mu^2 + \mu + 1\right)} \tag{11.24}$$

and - with $R_L \to \infty$ the gain $G_{wcf.b.nL}$ for the non-loaded output becomes:

$$G_{wcf.b.nL} = \frac{\mu\left(r_a + \mu R_{al.eff}\right)}{r_a\left(\mu + 2\right) + R_{al.eff}\left(\mu^2 + \mu + 1\right)} \tag{11.25}$$

11.4.1.2   Output resistance $R_{o.b}$:

At the gain $G_{wcf.b.L} = 0.5*G_{wcf.b.nL}$ the output resistance $R_{o.b}$ equals the load resistance R_L. Hence,

$$0.5\, G_{wcf.b.nL} = G_{wcf.b.L}$$

$$0.5 = \frac{\dfrac{\mu\left(r_a + \mu R_{al.eff}\right)}{\dfrac{r_a\left(r_a + R_{al.eff}\right)}{R_{o.b}} + r_a\left(\mu + 2\right) + R_{al.eff}\left(\mu^2 + \mu + 1\right)}}{\dfrac{\mu\left(r_a + \mu R_{al.eff}\right)}{r_a\left(\mu + 2\right) + R_{al.eff}\left(\mu^2 + \mu + 1\right)}} \tag{11.26}$$

rearrangement leads to $R_{o.b}$:

$$R_{o.b} = \frac{r_a\left(r_a + R_{a1.eff}\right)}{r_a\left(\mu + 2\right) + R_{a1.eff}\left(\mu^2 + \mu + 1\right)} \tag{11.27}$$

11.4.2   Un-bypassed version:

11.4.1.1   Gains $G_{wcf.u}$ (output loaded with R_L or non-loaded):

$$G_{wcf.u} = \frac{v_o}{v_i} \tag{11.28}$$

sum of currents at the top node:

$$i_o = i_1 + i_2 + i_3 - i_5 - i_6$$
$$\text{set } i_o = 0 \tag{11.29}$$
$$\Rightarrow i_1 = \frac{v_o}{R_L} = i_5 + i_6 - i_2 - i_3$$

hence,

$$\frac{v_o}{R_L} = g_m v_{g1.c1} + \frac{v_{a1.c1}}{r_a} - g_m v_{g2.c2} - \frac{v_{a2.c2}}{r_a} \tag{11.30}$$

with:

$$\begin{aligned} v_{a2.c2} &= v_o - v_{R.c2} \\ v_{g2.c2} &= -v_{R.a1.eff} - v_{R.c2} \\ v_{g1.c1} &= v_i - v_o \\ v_{a1.c1} &= -v_o - v_{R.a1.eff} \end{aligned} \tag{11.31}$$

and:

$$\begin{aligned} v_{R.c2} &= R_{c2}\left(i_2 + i_3\right) \\ v_{R.c2} &= R_{c2}\frac{v_o - v_{R.a1.eff}\,\mu}{\left(r_a + R_{c2} + \mu R_{c2}\right)} \end{aligned} \tag{11.32}$$

$$\begin{aligned} v_{R.a1.eff} &= i_7\, R_{a1.eff} \\ &= \left(i_5 + i_6\right) R_{a1.eff} \\ &= R_{a1.eff}\left(g_m v_{g1.c1} + \frac{v_{a1.c1}}{r_a}\right) \end{aligned} \tag{11.33}$$

we'll get:

$$v_{R.a1.eff} = R_{a1.eff} \frac{\mu v_i - (\mu + 1) v_o}{r_a + R_{a1.eff}} \tag{11.34}$$

and - after a lot of rearrangements - we'll get the gain $G_{wcf.u.L}$ that includes the output load R_L:

$$G_{wcf.u.L} = \frac{\mu r_a + \mu^2 R_{a1.eff} + (\mu^2 + \mu) R_{c2}}{\left(\dfrac{A}{R_L} + B\right)} \tag{11.35}$$

$$\begin{aligned} A &= r_a \left[r_a + R_{a1.eff} + (\mu + 1) R_{c2} \right] + (\mu + 1) R_{c2} R_{a1.eff} \\ B &= (\mu + 2) r_a + (\mu^2 + \mu + 1) R_{a1.eff} + (\mu^2 + 2\mu + 1) R_{c2} \end{aligned} \tag{11.36}$$

and - with $R_L \to \infty$ the gain $G_{wcf.b.nL}$ for the non-loaded output becomes:

$$G_{wcf.u.nL} = \frac{\mu r_a + \mu^2 R_{a1.eff} + (\mu^2 + \mu) R_{c2}}{(\mu + 2) r_a + (\mu^2 + \mu + 1) R_{a1.eff} + (\mu^2 + 2\mu + 1) R_{c2}} \tag{11.37}$$

11.4.1.2   Output resistance $R_{o.u}$:

At the gain $G_{wcf.u.L} = 0.5 * G_{wcf.u.nL}$ the output resistance $R_{o.u}$ equals the load resistance R_L. Hence,

$$0.5 G_{wcf.u.nL} = G_{wcf.u.L} \tag{11.38}$$

rearrangement leads to $R_{o.u}$:

$$R_{o.u} = \frac{r_a (r_a + R_{a1.eff}) \left[r_a + (\mu + 1) R_{c2} \right]}{(\mu + 2) r_a + (\mu^2 + \mu + 1) R_{a1.eff} + (\mu^2 + 2\mu + 1) R_{c2}} \tag{11.39}$$

11.5 Gain stage frequency and phase response calculations

I abstain from going through the calculation course to get the frequency and phase response. All these calculations can be derived from the respective paragraphs of the previous chapters.

As long as C_{c2}, C_{g2}, C_{in} and C_{out} are chosen of such values that do not hurt the flat frequency and phase response in B_{20k}, than, only the gain stage input frequency and phase response calculations with V1 related input capacitances are of further interest. Because of the rather low output resistance valve related output capacitances - other than the DC voltage blocking C_{out} - won't play a response flatness hurting role.

As could be demonstrated in the CAS gain stage chapter (6) the input capacitance (Miller-C!) of V2 in conjunction with the output resistance of V1 won't hurt the flat frequency and phase response in B_{20k} as well.

11.6   WCF example with V1,2 = E188CC / 7308
- bypassed (b) and un-bypassed (u) versions:

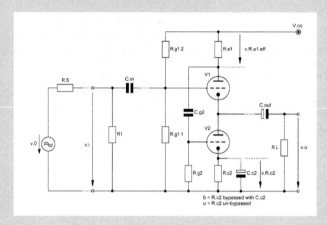

Figure 11.3 WCF example circuitry

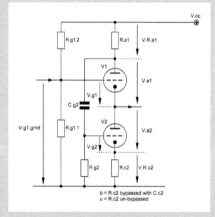

Figure 11.4 DC voltage overview

11.6.1 Triodes bias and other circuitry data:

$I_{a1} := 10 \cdot 10^{-3} A$ $I_{a2} := I_{a1}$ $V_{a1} := 90V$ $V_{a2} := 90V$

$V_{g1} := -1.65V$ $V_{g2} := -1.65V$

$V_{R.c2} := |V_{g2}|$

$R_{g2} := 1 \cdot 10^6 \Omega$ $R_{a1} := 12 \cdot 10^3 \Omega$

$R_{g1.1} := 1 \cdot 10^6 \Omega$ $R1 := 1 \cdot 10^6 \Omega$

➢ **MCD Worksheet XI:**   **WCF calculations**   **Page 2**

$$R_{c2} := \frac{V_{R.c2}}{I_{a2}}$$

$$R_{c2} = 165 \times 10^0 \, \Omega$$

$$R_{a1.eff} := \left(\frac{1}{R_{a1}} + \frac{1}{R_{g2}} \right)^{-1}$$

$$R_{a1.eff} = 11.858 \times 10^3 \, \Omega$$

$$V_{cc} := V_{a1} + V_{a2} + I_{a1} \cdot (R_{a1} + R_{c2})$$

$$V_{cc} = 301.65 \times 10^0 \, V$$

$$V_{R.a1} := I_{a1} \cdot R_{a1}$$

$$V_{R.a1} = 120 \times 10^0 \, V$$

$$V_{g1.grnd} := V_{cc} - V_{a1} - V_{R.a1} - |V_{g1}|$$

$$V_{g1.grnd} = 90 \times 10^0 \, V$$

$$R_{g1.2} := R_{g1.1} \cdot \frac{(V_{cc} - V_{g1.grnd})}{V_{g1.grnd}}$$

$$R_{g1.2} = 2.352 \times 10^6 \, \Omega$$

11.6.2. Triodes valve constants:

$$g_m := 10.4 \cdot 10^{-3} \, S \qquad r_a := 3.1 \cdot 10^3 \, \Omega \qquad \mu := g_m \cdot r_a \qquad g_m \cdot r_a = 32.24 \times 10^0$$

$$g_{m1} := g_m \qquad\qquad r_{a1} := r_a \qquad\qquad \mu_1 := \mu$$

$$g_{m2} := g_m \qquad\qquad r_{a2} := r_a \qquad\qquad \mu_2 := \mu$$

11.6.3 Gains $G_{wcf.b.L}$ and $G_{wcf.u.L}$:

$$R_L := 10 \cdot 10^3 \, \Omega$$

$$G_{wcf.b.L} := \frac{\mu \cdot (r_a + \mu \cdot R_{a1.eff})}{\dfrac{r_a^2 + r_a \cdot R_{a1.eff}}{R_L} + r_a \cdot (\mu + 2) + R_{a1.eff} (\mu^2 + \mu + 1)}$$

$$G_{wcf.b.L} = 968.434 \times 10^{-3}$$

$$A := r_a \left[r_a + R_{a1.eff} + (\mu + 1) \cdot R_{c2} \right] + (\mu + 1) \cdot R_{c2} \cdot R_{a1.eff}$$

$$B := (\mu + 2) \cdot r_a + (\mu^2 + 2\mu + 1) \cdot R_{c2} + (\mu^2 + \mu + 1) \cdot R_{a1.eff}$$

$$G_{wcf.u.L} := \frac{\mu \cdot (r_a + \mu \cdot R_{a1.eff}) + (\mu^2 + \mu) \cdot R_{c2}}{\dfrac{A}{R_L} + B}$$

$$G_{wcf.u.L} = 967.845 \times 10^{-3}$$

$$R_L := 1.1\Omega, 1.2\Omega \, .. \, 100\Omega$$

$$G_{wcf.b.L}(R_L) := \frac{\mu \cdot (r_a + \mu \cdot R_{a1.eff})}{\dfrac{r_a^2 + r_a \cdot R_{a1.eff}}{R_L} + r_a \cdot (\mu + 2) + R_{a1.eff} (\mu^2 + \mu + 1)}$$

➤ **MCD Worksheet XI:**   **WCF calculations**

$$G_{wcf.u.l}(R_L) := \frac{\mu \cdot (r_a + \mu \cdot R_{a1.eff}) + (\mu^2 + \mu) \cdot R_{c2}}{\dfrac{A}{R_L} + B}$$

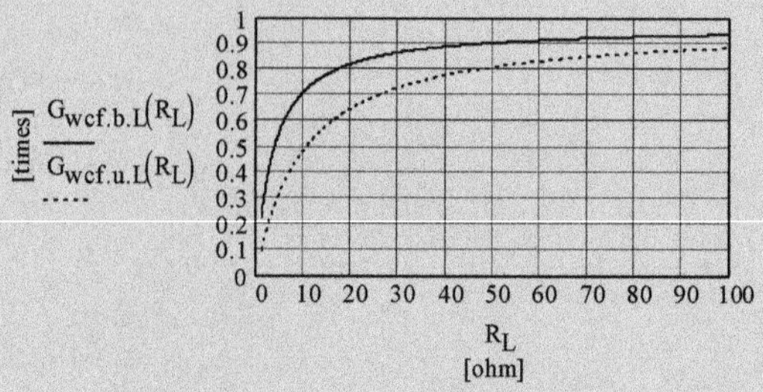

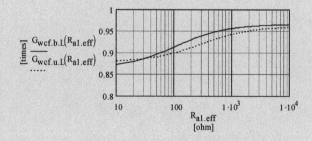

$$\begin{array}{c} \underset{[times]}{} \dfrac{G_{wcf.b.l}(R_L)}{G_{wcf.u.l}(R_L)} \\ \text{-----} \end{array}$$

R_L
[ohm]

Figure 11.5 Gains $G_{wcf.b.l}$ and $G_{wcf.u.l}$ vs. R_L with fixed $R_{a1.eff}$

$$R_{a1.eff} := 10\Omega, 20\Omega .. 10000\,\Omega \qquad\qquad\qquad R_L := 1 \cdot 10^3 \Omega$$

$$G_{wcf.b.l}(R_{a1.eff}) := \frac{\mu \cdot (r_a + \mu \cdot R_{a1.eff})}{\dfrac{r_a^2 + r_a \cdot R_{a1.eff}}{R_L} + r_a \cdot (\mu + 2) + R_{a1.eff}(\mu^2 + \mu + 1)}$$

$$A(R_{a1.eff}) := r_a \cdot \left[r_a + R_{a1.eff} + (\mu + 1) \cdot R_{c2} \right] + (\mu + 1) \cdot R_{c2} \cdot R_{a1.eff}$$

$$B(R_{a1.eff}) := (\mu + 2) \cdot r_a + (\mu^2 + 2\mu + 1) \cdot R_{c2} + (\mu^2 + \mu + 1) \cdot R_{a1.eff}$$

$$G_{wcf.u.l}(R_{a1.eff}) := \frac{\mu \cdot (r_a + \mu \cdot R_{a1.eff}) + (\mu^2 + \mu) \cdot R_{c2}}{\dfrac{A(R_{a1.eff})}{R_L} + B(R_{a1.eff})}$$

$$\begin{array}{c} \underset{[times]}{} \dfrac{G_{wcf.b.l}(R_{a1.eff})}{G_{wcf.u.l}(R_{a1.eff})} \\ \text{-----} \end{array}$$

$R_{a1.eff}$
[ohm]

Figure 11.6 Gains $G_{wcf.b.l}$ and $G_{wcf.u.l}$ vs. $R_{a1.eff}$ with fixed R_L

11.6.4 Gains $G_{wcf.b.nL}$ and $G_{wcf.u.nL}$:

$$R_{a1.eff} := \left(\frac{1}{R_{a1}} + \frac{1}{R_{g2}}\right)^{-1}$$

$$G_{wcf.b.nL} := \frac{\mu \cdot (r_a + \mu \cdot R_{a1.eff})}{r_a \cdot (\mu + 2) + R_{a1.eff}(\mu^2 + \mu + 1)} \qquad\qquad G_{wcf.b.nL} = 968.785 \times 10^{-3}$$

$$G_{wcf.u.nL} := \frac{\mu \cdot r_a + \mu^2 \cdot R_{a1.eff} + (\mu^2 + \mu) \cdot R_{c2}}{(\mu + 2) \cdot r_a + (\mu^2 + 2\mu + 1) \cdot R_{c2} + (\mu^2 + \mu + 1) \cdot R_{a1.eff}} \qquad G_{wcf.u.nL} = 968.8004 \times 10^{-3}$$

$$R_{a1.eff} := 10\Omega, 20\Omega .. 10000\Omega$$

$$G_{wcf.b.nL}(R_{a1.eff}) := \frac{\mu \cdot (r_a + \mu \cdot R_{a1.eff})}{r_a \cdot (\mu + 2) + R_{a1.eff}(\mu^2 + \mu + 1)}$$

$$G_{wcf.u.nL}(R_{a1.eff}) := \frac{\mu \cdot r_a + \mu^2 \cdot R_{a1.eff} + (\mu^2 + \mu) \cdot R_{c2}}{(\mu + 2) \cdot r_a + (\mu^2 + 2\mu + 1) \cdot R_{c2} + (\mu^2 + \mu + 1) \cdot R_{a1.eff}}$$

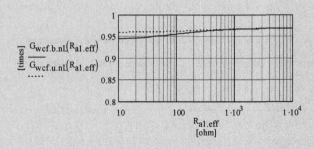

Figure 11.7 Gains $G_{wcf.b.nL}$ and $G_{wcf.u.nL}$ vs. $R_{a1.eff}$

11.6.5 Output resistances $R_{o.b}$ and $R_{o.u}$:

$$R_{a1.eff} := \left(\frac{1}{R_{a1}} + \frac{1}{R_{g2}}\right)^{-1}$$

$$R_{o.b} := \frac{r_a \cdot (r_a + R_{a1.eff})}{(\mu + 2) \cdot r_a + (\mu^2 + \mu + 1) \cdot R_{a1.eff}} \qquad\qquad R_{o.b} = 3.615 \times 10^0 \Omega$$

$$R_{o.b.rot} := \frac{r_a}{\mu \cdot R_{a1}} \cdot \frac{R_{a1} + r_a}{1 + \mu} \qquad\qquad R_{o.b.rot} = 3.64 \times 10^0 \, \Omega$$

$$R_{a1} := 100\,\Omega, 110\,\Omega \, .. \, 10000\,\Omega$$

$$R_{o.b}(R_{a1}) := r_a \cdot \frac{r_a + \left(\dfrac{1}{R_{a1}} + \dfrac{1}{R_{g2}}\right)^{-1}}{(\mu + 2) \cdot r_a + \left(\mu^2 + \mu + 1\right) \cdot \left(\dfrac{1}{R_{a1}} + \dfrac{1}{R_{g2}}\right)^{-1}}$$

$$R_{o.b.rot}(R_{a1}) := \frac{r_a}{\mu \cdot R_{a1}} \cdot \frac{R_{a1} + r_a}{1 + \mu}$$

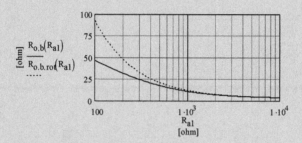

Figure 11.8 Comparison between exact and rule-o-thumb $R_{o.b}$ calculation

Conclusion: for an error <10% calculation of $R_{o.b.rot}$ makes sense for R_{a1} > 1k only

$$R_{o.u} := \frac{\left(r_a + R_{a1.eff}\right) \cdot \left[r_a + (\mu + 1) \cdot R_{c2}\right]}{(\mu + 2) \cdot r_a + \left(\mu^2 + 2\mu + 1\right) \cdot R_{c2} + \left(\mu^2 + \mu + 1\right) \cdot R_{a1.eff}} \qquad R_{o.u} = 9.872 \times 10^0 \, \Omega$$

$$R_{a1.eff} := 10\,\Omega, 20\,\Omega \, .. \, 10000\,\Omega$$

$$R_{o.b}(R_{a1.eff}) := \frac{r_a \cdot \left(r_a + R_{a1.eff}\right)}{(\mu + 2) \cdot r_a + \left(\mu^2 + \mu + 1\right) \cdot R_{a1.eff}}$$

$$R_{o.u}(R_{a1.eff}) := \frac{\left(r_a + R_{a1.eff}\right) \cdot \left[r_a + (\mu + 1) \cdot R_{c2}\right]}{(\mu + 2) \cdot r_a + \left(\mu^2 + 2\mu + 1\right) \cdot R_{c2} + \left(\mu^2 + \mu + 1\right) \cdot R_{a1.eff}}$$

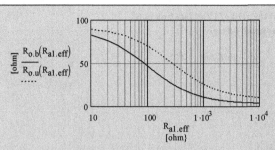

Figure 11.9 $R_{o.b}$ and $R_{o.u}$ vs. $R_{al.eff}$

11.6.6 Specific resistances and capacitances:

$$R_{al} := 12 \cdot 10^3 \Omega$$

$$R_{al.eff} := \left(\frac{1}{R_{al}} + \frac{1}{R_{g2}} \right)^{-1}$$

$$R_i := \left(\frac{1}{R1} + \frac{1}{R_{g1.1}} + \frac{1}{R_{g1.2}} \right)^{-1} \qquad\qquad R_i = 412.332 \times 10^3 \Omega$$

$$f_{c.opt} := 0.2 \text{Hz}$$

$$C_{in} := \frac{1}{2 \cdot \pi \cdot f_{c.opt} \cdot \left(\frac{1}{R_{g1.1}} + \frac{1}{R_{g1.2}} \right)^{-1}} \qquad\qquad C_{in} = 1.134 \times 10^{-6} \text{F}$$

$$R_L := 10 \cdot 10^3 \Omega$$

$$C_{out} := \frac{1}{2 \cdot \pi \cdot f_{c.opt} \cdot R_L} \qquad\qquad C_{out} = 79.577 \times 10^{-6} \text{F}$$

$$C_{g2} := \frac{1}{2 \cdot \pi \cdot f_{c.opt} \cdot R_{g2}} \qquad\qquad C_{g2} = 795.775 \times 10^{-9} \text{F}$$

$$r_{c2} := \frac{R_{al.eff} + 2 \cdot r_a}{\mu + 1} \qquad\qquad r_{c2} = 543.252 \times 10^0 \Omega$$

$$C_{c2} := \frac{1}{2 \cdot \pi \cdot f_{c.opt} \cdot \left(\frac{1}{r_{c2}} + \frac{1}{R_{c2}} \right)^{-1}} \qquad\qquad C_{c2} = 6.288 \times 10^{-3} \text{F}$$

➢ MCD Worksheet XI: WCF calculations Page 7

11.6.7 Voltage $v_{R.a1.eff}$ at the grid of V2 as function of R_L for a specific rms input voltage v_i:

$$R_{a1.eff} := 10 \cdot 10^3 \, \Omega \qquad\qquad R_L := 10\Omega, 20\Omega .. 10000\Omega$$

$$v_i := 1V \qquad\qquad v_{i.b.p} := v_i \sqrt{2} \qquad\qquad v_{i.b.p} = 1.414 \times 10^0 \, V$$

$$v_{i.u.p} := v_{i.b.p}$$

$$v_{o.b.p}(R_L) := v_{i.b.p} \cdot G_{wcf.b.L}(R_L)$$

$$v_{o.u.p}(R_L) := v_{i.u.p} \cdot G_{wcf.u.L}(R_L) \qquad\qquad v_{R.a1.eff} = R_{a1.eff} \frac{\mu \cdot v_i - v_o \cdot (1+\mu)}{r_a + R_{a1.eff}}$$

$$v_{b.R.a1.eff.p}(R_L) := R_{a1.eff} \frac{\mu \cdot v_{i.b.p} - v_{o.b.p}(R_L) \cdot (1+\mu)}{r_a + R_{a1.eff}}$$

$$v_{u.R.a1.eff.p}(R_L) := R_{a1.eff} \frac{\mu \cdot v_{i.u.p} - v_{o.u.p}(R_L) \cdot (1+\mu)}{r_a + R_{a1.eff}}$$

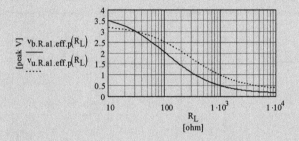

Figure 11.10 V2 grid peak AC voltage vs. R_L

Conclusions for 90% $|V_{g2}|$: (b) version: $R_{L.b}$ should be >190R

(u) version: $R_{L.u}$ should be >430R

11.6.8 output current check (should be < app. 90% of I_{a1}):

$$R_{L.b} := 190\Omega \qquad\qquad R_{L.u} := 430\Omega$$

$$i_{out.b.p} := \frac{v_{o.b.p}(R_{L.b})}{R_{L.b}} \qquad\qquad i_{out.b.p} = 6.915 \times 10^{-3} \, A$$

$$i_{out.u.p} := \frac{v_{o.u.p}(R_{L.u})}{R_{L.u}} \qquad\qquad i_{out.u.p} = 3.052 \times 10^{-3} \, A$$

Chapter 12 Triodes in Parallel Operation (PAR):

12.1 Circuit diagram

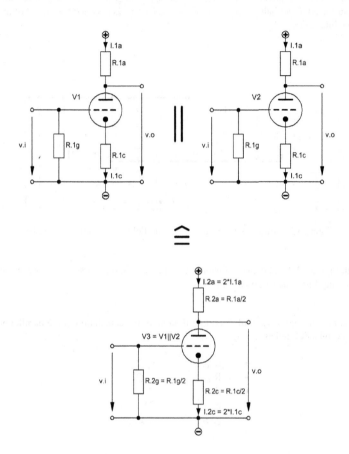

Figure 12.1 Two triodes in parallel operation

12.2 Equivalent circuit plus resulting component and valve figures

It is obvious that the DC current conditions of two equally configured triodes (top 2/3 of Figure 12.1) must get doubled when putting them in parallel operation à la bottom 1/3 of Figure 12.1. Assumed that the power supply voltages do not change plate and cathode resistors must be halved. While there is practically no grid current the grid resistor R_{2g} doesn't need to get halved. It can be set to a value that is right for the output of the preceding stage.

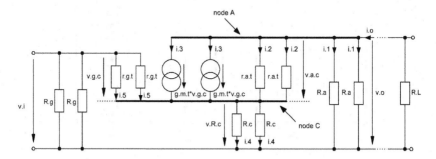

Figure 12.2 Equivalent circuit of the parallel operation of two triodes

With Figure 12.2 and the comprehension of the derivation course of Chapter 1 the following formulae can be developed.

According to Figure 12.2 all important figures for the calculation of the parallel operation ("$_2$") can be defined as:

- $r_{a.2t}$ $= r_{a.t}/2$
- R_{2a} $= R_a/2$
- R_{2c} $= R_c/2$
- $g_{m.2t}{}^*v_{g.c}$ $= 2*g_{m.t}{}^*v_{g.c}$
- $g_{m.2t}$ $= 2*g_{m.t}$
- μ_{2t} $= \mu_t$

12.3 Basic formulae (excl. stage load (R_L)-effect)

12.3.1 Gain G_{2t} of the new triode in parallel operation

becomes in terms of $g_{m.t}$:

$$G_{2t} = -g_{m.2t} \frac{r_{a.2t} R_{2a}}{r_{a.2t} + R_{2a} + R_{2c}\left(1 + g_{m.2t}r_{a.2t}\right)} \tag{12.1}$$

becomes in terms of μ_t:

$$G_{2t} = -\mu_t \frac{R_{2a}}{r_{a.2t} + R_{2a} + R_{2c}\left(1 + \mu_t\right)} \tag{12.2}$$

thus, compared with the single triode operation the gain of the parallel operation does not change

12.3.2 Grid input impedance $Z_{i.2g.eff}$:

$$Z_{i.2g.eff} = R_g \parallel C_{2i.tot} \tag{12.3}$$

12.3.3 Plate output impedance $Z_{o.2a.eff}$:

$$Z_{o.2a.eff} = R_{2a} \parallel \left[r_{a.2t} + \left(1 + \mu_t\right)R_{2c}\right] \parallel \left(C_{2o}\right) \tag{12.4}$$

this is 1/2 of the respective figure of the single triode gain stage

12.3.4 Cathode output resistance $R_{o.2c}$:

$$R_{o.2c} = R_{2c} \parallel \frac{R_{2a} + r_{a.2t}}{1 + \mu_t} \tag{12.5}$$

this is 1/2 of the respective figure of the single triode gain stage

12.3.5 Total input capacitance $C_{2i.tot}$ and output capacitance $C_{2o.tot}$:

- with plate-cathode capacitance $C_{a.c.1t}$
- with grid-cathode capacitance $C_{g.c.1t}$ for one triode
- with grid-plate capacitance $C_{g.a.1t}$ for one triode
- with grid input Miller capacitance effect $[(1+|G_{1t}|)*C_{g.a.1t}]$ for one triode
- with plate output Miller capacitance effect $[C_{a.c.1t} \parallel C_{g.a.1t}]$ for one triode
- with a guessed stray capacitance $C_{stray.1t}$ for 1 triode

the total input capacitance for two paralleled triodes becomes two times the total input capacitance for one triode:

$$C_{2i.tot} = 2\left[\left(1 + |G_{2t}|\right)C_{g.a.1t} + C_{g.c.1t} + C_{stray.1t}\right] \tag{12.6}$$

the respective output capacitance C_{2o} becomes

$$C_{2o.tot} = 2\,C_o = 2\left(C_{a.c.1t} + C_{g.a.1t}\right) \tag{12.7}$$

12.3.6   Bypassing the cathode resistor with a capacitor means that in the above given
equations the term R_{2c} has to be set to zero. The capacitor should be of a size[1] that
does not hurt a flat frequency and phase response in B_{20k}.

12.4 n triodes in parallel operation

12.4.1   According to Figure 12.2 all important figures for the calculation of the
parallel operation ("$_n$") can be defined as:

- $r_{a.nt}$ $= r_{a.t}/n$
- R_{na} $= R_a/n$
- R_{nc} $= R_c/n$
- $g_{m.nt}{}^*v_{g.c}$ $= n^*g_{m.t}{}^*v_{g.c}$
- $g_{m.nt}$ $= n^*g_{m.t}$
- μ_{nt} $= \mu_t$

12.4.2   The gain G_{nt} of the new triode in parallel operation

becomes in terms of $g_{m.t}$:

$$G_{nt} = -\,g_{m.nt}\,\frac{r_{a.nt}\,R_{na}}{r_{a.nt} + R_{na} + R_{nc}\left(1 + g_{m.nt}r_{a.nt}\right)} \tag{12.8}$$

becomes in terms of μ_t:

$$G_{nt} = -\,\mu_t\,\frac{R_{na}}{r_{a.nt} + R_{na} + R_{nc}\left(1 + \mu_t\right)} \tag{12.9}$$

thus, compared with the single triode operation the gain of the parallel operation
does not change

12.4.3   Grid input impedance $Z_{i.ng.eff}$:

$$Z_{i.ng.eff} = R_g \parallel C_{ni.tot} \tag{12.10}$$

---

[1] see Chapter 2

12.4.4   Plate output impedance $Z_{o.na.eff}$:

$$Z_{o.na.eff} = R_{na} \parallel \left[r_{a.nt} + (1+\mu_t) R_{nc} \right] \parallel C_{no} \qquad (12.11)$$

this is 1/n of the respective figure of the single triode gain stage

12.4.5   Cathode output resistance $R_{o.nc}$:

$$R_{o.nc} = R_{nc} \parallel \frac{R_{na} + r_{a.nt}}{1+\mu_t} \qquad (12.12)$$

this is 1/n of the respective figure of the single triode gain stage

12.4.6   Total input capacitance $C_{ni.tot}$ and output capacitance $C_{no.tot}$:

$$C_{ni.tot} = n \left[(1+|G_{nt}|) C_{g.a.1t} + C_{g.c.1t} + C_{stray.1t} \right] \qquad (12.13)$$

$$C_{no.tot} = n C_o = n \left(C_{a.c.1t} + C_{g.a.1t} \right) \qquad (12.14)$$

12.4.7   Bypassing the cathode resistor with a capacitor means that in the above given equations the term R_{nc} has to be set to zero. The capacitor should be of a size[2] that does not hurt a flat frequency and phase response in B_{20k}.

---

[2] see Chapter 2

12.5 Advantages of the parallel operation of n triodes

- in case of a CF configuration: decreasing output impedance:
                                      multiplication with factor 1/n

- in case of a typical gain stage: deacreasing noise voltage:
                                      multiplication with factor $(1/\sqrt{n})$

12.6 Disadvantages of the parallel operation of n triodes

- growing current consumption (n times)

- total input capacitance $C_{ni.tot}$ grows with factor n

- total output capacitance $C_{no.tot}$ grows with factor n

12.7 Gain stage frequency and phase response calculations

- In consideration of the number of triodes in parallel operation all these calculations
  can be derived from the respective paragraphs of the previous chapters[3].

---

[3]   see also Chapter 15 "Design Example"

Chapter 13 Balanced (Differential) Gain Stage (BAL)

13.1 Circuit diagram

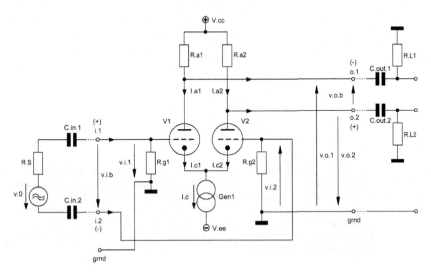

Figure 13.1 Two triodes configured as a balanced (differential) gain stage
for AC signal amplification purposes

13.2   Basic assumptions

To ensure correct work of a differential gain stage the following assumption have to be made:

$$R_a = R_{a1} = R_{a2}$$
$$R_g = R_{g1} = R_{g2} \tag{13.1}$$

$$V1 = V2 = double\ triode$$
$$\mu_t = \mu_1 = \mu_2$$
$$g_{m.t} = g_{m.1} = g_{m.2} \tag{13.2}$$
$$r_{a.t} = r_{a.1} = r_{a.2}$$

$$I_c = I_{c1} + I_{c2} = I_{a1} + I_{a2} \tag{13.3}$$

DC current I_c keeps constant in any case of changing input voltages, that means that a change of v_{i1} creates a certain change of I_{a1}, thus, creating exactly the same amount of current change of I_{a2} with the opposite polarity. Therefore, in the equivalent circuit environment Gen1 works as short circuit between cathode and the ground level of the circuitry. Hence, as of Figure 13.2 V1 and V2 can be treated as two CCS gain stages with grounded cathodes, each amplifying half of the differential input voltage.

13.3   Amplification variants

With the generic gain stage of Figure 13.1 and the phase polarization of the in- and output voltages as indicated the respective gains G_{xyz} and the following amplification variants can be chosen:

13.3.1   Gain for balanced (b) input via i.1 and i.2 – balanced output via o.1 and o.2:

$$G_{b.b} = \frac{v_{o.b}}{v_{i.b}} \tag{13.4}$$

$$G_{b.b} = -g_{m.t} \left(r_{a.t} \parallel R_a \right) \tag{13.5}$$

13.3.2   Gain for balanced input via i.1 and i.2 – unbalanced (u) output via o.1 or o.2:

13.3.2.1   with 180° phase shift via output o.1 only:

$$G_{b.u.o1} = \frac{v_{o.1}}{v_{i.b}} = -\frac{1}{2} g_{m.t} \left(r_{a.t} \parallel R_a \right) \tag{13.6}$$

13.3.2.2   without phase shift via output o.2 only:

$$G_{b.u.o2} = \frac{v_{o.2}}{v_{i.b}} = \frac{1}{2} g_{m.t} \left(r_{a.t} \parallel R_a \right) \tag{13.7}$$

13.3.3   Gain for unbalanced input via input i.1 and input i.2 grounded – balanced output via o.1 and o.2:

$$G_{u.i1.b} = \frac{v_{o.b}}{v_{i.1}} = -2 g_{m.t} \left(r_{a.t} \parallel R_a \right) \tag{13.8}$$

13.3.4   Gain for unbalanced input via input i.2 and input i.1 grounded – balanced output via o.1 and o.2:

$$G_{u.i2.b} = \frac{v_{o.b}}{v_{i.2}} = 2 g_{m.t} \left(r_{a.t} \parallel R_a \right) \tag{13.9}$$

13.3.5   Gain for unbalanced input via i.1 and input i.2 grounded – unbalanced output via o.1 only:

$$G_{u.i1.u.o1} = \frac{v_{o.1}}{v_{i.1}} = -g_{m.t} \left(r_{a.t} \parallel R_a \right) \tag{13.10}$$

13.3.6   Gain for unbalanced input via i.1 and input i.2 grounded – unbalanced output via o.2 only:

$$G_{u.i1.u.o2} = \frac{v_{o.2}}{v_{i.1}} = g_{m.t} \left(r_{a.t} \parallel R_a \right) \tag{13.11}$$

13.3.7   Gain for unbalanced input via i.2 and input i.1 grounded – unbalanced output via o.2 only:

$$G_{u.i2.u.o2} = \frac{v_{o.2}}{v_{i.2}} = -g_{m.t} \left(r_{a.t} \parallel R_a \right) \tag{13.12}$$

13.3.8   Gain for unbalanced input via i.2 and input i.1 grounded – unbalanced output via o.1 only:

$$G_{u.i2.u.o1} = \frac{v_{o.1}}{v_{i.2}} = g_{m.t} \left(r_{a.t} \parallel R_a \right) \tag{13.13}$$

13.4   Derivations

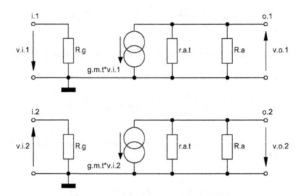

Figure 13.2 Simplified equivalent circuits for each part
of the differential (balanced) input voltage $v_{i.b}$ of Figure 13.1

Differential gain $G_{b.b}$ for a differential input voltage $v_{i.b}$:

$$G_{b.b} = \frac{v_{o.b}}{v_{i.b}} \qquad (13.14)$$

gain G_{v1} of V1 and gain G_{v2} of V2:

$$v_{i.b} = v_{i.1} - v_{i.2} \qquad (13.15)$$

$$v_{i.2} = - v_{i.1} \qquad (13.16)$$

$$G_{v1} = -\frac{v_{o.1}}{v_{i.b}}$$

$$G_{v2} = \frac{v_{o.2}}{v_{i.b}} \qquad (13.17)$$

$$G_t = G_{v1} = - G_{v2} = -\frac{1}{2}\mu_t \frac{R_a}{r_a + R_a}$$

$$G_t = -\frac{1}{2}g_{m.t}\left(r_{a.t} \parallel R_a\right) \qquad (13.18)$$

thus, $G_{b.b}$ becomes:

$$v_{o.1} = - G_t\, v_{i.b}$$

$$v_{o.2} = G_t\, v_{i.b}$$

$$v_{o.b} = v_{o.1} - v_{o.2} \qquad (13.19)$$

$$\Rightarrow G_{b.b} = \frac{v_{o.b}}{v_{i.b}} = -2G_t = -g_{m.t}\left(r_{a.t} \parallel R_a\right) \tag{13.20}$$

All other gain expressions are included in the above given equations.

13.5   Impedances

13.5.1   Balanced input impedance $Z_{i.b}$ between i.1 and i.2 (C_i includes Miller-C of each triode):

$$Z_{i.b} = \left(2R_g\right) \parallel \left(0.5 C_i\right) \tag{13.21}$$

$$C_i = C_{i.v1} = C_{i.v2} = \left(1 + G_t\right)C_{g.a} + C_{g.c} + C_{stray} \tag{13.22}$$

13.5.2   Unbalanced input impedance $Z_{i.u}$ between i.1 or i.2 and ground

$$Z_{i.u} = R_g \parallel C_i \tag{13.23}$$

13.5.3   Balanced output impedance $Z_{o.b}$ between o.1 and o.2:

$$Z_{o.b} = \left[2\left(r_a \parallel R_a\right)\right] \parallel \left(0.5 C_o\right) \tag{13.24}$$

$$C_o = C_{a.c} \tag{13.25}$$

13.5.4   Unbalanced output impedance $Z_{o.u}$ between o.1 or o.2 and ground:

$$Z_{o.u} = \left(r_a \parallel R_a\right) \parallel C_o \tag{13.26}$$

13.6   Common mode gain G_{com} for a common mode input voltage $v_{i.com}$

Common mode means that with reference to the ground level of the circuitry the input voltages at i1 and i2 are totally equal concerning phase, frequency and amplitude. The respective equivalent circuit looks as follows:

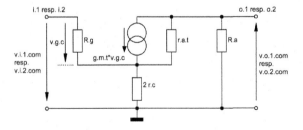

Figure 13.3 Equivalent circuit for the common mode
input voltage situation of V1 and V2 respectively

$$v_{i.com} = v_{i.1.com} + v_{i.2.com} \tag{13.27}$$

The common mode gain $G1_{com}$ becomes the gain of a CCS gain stage with a cathode resistance of 2 times the internal resistance of the current generator Gen1 of Figure 13.1 (see Figure 13.4):

$$G1_{com} = \frac{v_{o.1.com}}{v_{i.com}} = -\frac{1}{2}\mu_t \frac{R_a}{r_{a.t} + R_a + (1 + \mu_t)2r_c} \tag{13.28}$$

$$\begin{aligned} G2_{com} &= G1_{com} \\ \Rightarrow G_{com} &= G1_{com} + G2_{com} \end{aligned} \tag{13.29}$$

$$G_{com} = -\mu_t \frac{R_a}{r_{a.t} + R_a + (1 + \mu_t)2r_c} \tag{13.30}$$

rule of thumb for G_{com}:

$$G_{com} = -\frac{R_a}{2r_c} \tag{13.31}$$

The derivation of the cathode resistance $2*r_c$ via equivalent circuit modelling looks as follows:

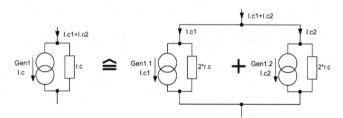

Figure 13.4 Derivation of the term "$2*r_c$"

13.7 CMRR - Common mode rejection ratio

13.7.1 Derivation:

$$CMRR = \frac{G_{b.b}}{G_{com}} \tag{13.32}$$

with equations (13.20) and (13.30) CMRR becomes:

$$CMRR = 1 + \frac{2r_c}{r_{a.t} + R_a} + \mu_t \frac{2r_c}{r_{a.t} + R_a} \tag{13.33}$$

$$CMRR_e = 20\log(CMRR) \tag{13.34}$$

Rule of thumb:

since

$$\mu_t \frac{2r_c}{r_{a.t} + R_a} \gg \frac{2r_c}{r_{a.t} + R_a} \gg 1 \tag{13.35}$$

and

$$r_{a.t} \approx R_a \tag{13.36}$$

CMRR approximately becomes (without big error):

$$CMRR_{rot} \approx \mu_t \frac{2r_c}{r_{a.t} + R_a} \tag{13.37}$$

$$CMRR_{rot} \approx g_{m.t}\, r_c \tag{13.38}$$

$$CMRR_{rot.e} = 20\log(CMRR_{rot}) \tag{13.39}$$

13.7.2 Sources of friction:

In theory - with an excellent current generator at the cathode - CMRR should become rather high margins. To achieve this the following facts are the real challenging ones:

- difference in the plate resistances R_{a1} and R_{a2}
- rather low value of the current generator's internal resistance r_c
- rather low value of $g_{m.t}$
- not perfectly matched valve gain μ_t
- differences in the valve capacitances

Low frequency ($\leq$ 1kHz) gain differences of V1 vs. V2 can be balanced by the
inclusion of P1 as shown in the following Figure 13.5 - or by its alternative
around P2:

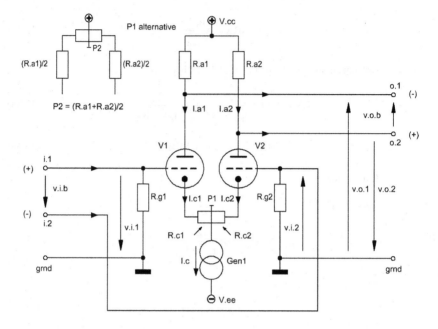

Figure 13.5 Improvements of Figure 13.1 to get
equal gains of V1 and V2

A disadvantage of the Figure 13.5 P1 approach is the fact that the mutual
conductance $g_{m.t}$ got reduced to a lower sized $g_{m.t.red}$, thus, changing $r_{a.t}$ as well,
hence, with a constant μ_t the respective figures for V1 will become:

$$g_{m.1.red} = \frac{g_{m.1}}{1 + g_{m.1}R_{c1}} \qquad (13.40)$$

$$r_{a.1.red} = \frac{\mu_t}{g_{m.1.red}} \qquad (13.41)$$

V2 has to be treated the same way! Depending on the size of R_{c1} and R_{c2} (should be
as small as possible) the result will be a smaller gain of the stage. P2 doesn't
produce these problems.

In any case, to sum up the CMRR story: because of the many unbalanced
components / active device values in a balanced valve gain stage it's a rather heavy
task to exactly calculate CMRR.

13.7.3   One additional question keeps open:

Despite the gain reduction by a reduced $g_{m.t}$: how to increase the gain of a differential gain stage?

Answer: Replacement of the plate resistors $R_{a.1}$ and $R_{a.2}$ by an active approach with valves that are configured as current generators, eg. like the top valves in the μ-Follower circuitry or as gain producing top devices like the ones in the cascoded gain stage will lead to higher gains and - in case of the μ-Follower only - to lower output impedances. The gain and impedance calculation courses are described in the respective chapters.

Taking into account the tiny gain reduction of the top section of a SRPP gain stage this would also work to produce lower output impedances - but with a bit less potential of overall gain increase.

It must be pointed out that, because of the inclusion of a cathode resistance, in any gain calculation V1 and V2 must be treated like CCS gain stages, thus $G_{b.b.red}$ becomes:

$$G_{b.b.red} = -\left(\begin{array}{c} \dfrac{1}{2} g_{m.1.red} \dfrac{r_{a.1.red} R_a}{r_{a.1.red} + R_a + (1+\mu_t) R_{c1}} \\[3mm] + \dfrac{1}{2} g_{m.2.red} \dfrac{r_{a.2} R_a}{r_{a.2.red} + R_a + (1+\mu_t) R_{c2}} \end{array} \right) \tag{13.42}$$

13.7.4   Rule of thumb:

With

$$R_{c1} \approx R_{c2}$$

$$r_{a.1.red} = r_{a.2.red} = r_{a.t.red} \tag{13.43}$$

$G_{b.b.red}$ becomes the following rule of thumb:

$$G_{b.b.red.rot} \approx - g_{m.t.red} \dfrac{r_{a.t.red} R_a}{r_{a.t.red} + R_a + (1+\mu_t) 0.5 \times P1} \tag{13.44}$$

13.8   Example with V1, V2 = E188CC / 7308 (188) and a gain of app. 10 (20dB):

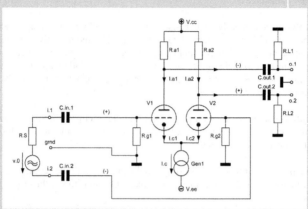

Figure 13.6 BAL gain stage example circuitry

13.8.1 Triode bias data:

$$I_{a1} = I_{a2} = 5 \cdot 10^{-3} A \qquad\qquad V_{a1} = 90V \qquad\qquad V_{a2} = 90V$$

$$V_{cc} = 103V \qquad V_{ee} := -15V \qquad V_{g1} = -2.1V \qquad V_{g2} = -2.1V$$

It is assumed that the design of the current generator Gen.1 is based on solid state components

13.8.2 Triode valve constants:

$$g_{m.188} := 7 \cdot 10^{-3} \cdot S \qquad \mu_{188} := 30.8 \qquad r_{a.188} := 4.4 \cdot 10^{3} \Omega$$

$$C_{g.c.1.188} := 3.1 \cdot 10^{-12} F \qquad C_{g.a.1.188} := 1.4 \cdot 10^{-12} F \qquad C_{a.c.1.188} := 1.75 \cdot 10^{-12} F$$

$$C_{g.c.2.188} := C_{g.c.1.188} \qquad C_{g.a.2.188} := C_{g.a.1.188} \qquad C_{a.c.2.188} := C_{a.c.1.188}$$

13.8.3 Circuit variables:

$$R_{a1} := 2.21 \cdot 10^{3} \Omega \qquad R_{a2} := R_{a1} \qquad R_{a} := R_{a1}$$

$$R_{g1} := 1 \cdot 10^{6} \Omega \qquad R_{g2} := R_{g1} \qquad R_{g} := R_{g1}$$

$$R_{gg1} := 0.1 \cdot 10^{3} \Omega \qquad R_{gg2} := R_{gg1} \qquad R_{gg} := R_{gg1}$$

$$R_{S} := 1 \cdot 10^{3} \Omega \qquad R_{L1} := 0.475 \cdot 10^{6} \Omega \qquad R_{L2} := R_{L1}$$

$$C_{stray.1.188} := 10 \cdot 10^{-12} F \qquad C_{in.1} := 1 \cdot 10^{-6} F \qquad C_{out.1} := 1 \cdot 10^{-6} F$$

$$C_{stray.2.188} := C_{stray.1.188} \qquad C_{in.2} := C_{in.1} \qquad C_{out.2} := C_{out.1}$$

13.8.4 Calculation relevant data:

frequency range f for the below shown graphs:

$$f := 10Hz, 20Hz .. 20000\,Hz$$

$$h := 1000 \cdot Hz$$

13.8.5 Gain variants:

$$G_{b.b} := -g_{m.188}\left(\frac{1}{r_{a.188}} + \frac{1}{R_a}\right)^{-1} \qquad G_{b.b} = -10.298 \times 10^0$$

$$G_{b.b.e} := 20 \cdot \log\left(|G_{b.b}|\right) \qquad G_{b.b.e} = 20.255 \times 10^0 \quad [dB]$$

$$G_{b.u.o1} := -\frac{1}{2} \cdot g_{m.188}\left(\frac{1}{r_{a.188}} + \frac{1}{R_a}\right)^{-1} \qquad G_{b.u.o1} = -5.149 \times 10^0$$

$$G_{b.u.o1.e} := 20 \cdot \log\left(|G_{b.u.o1}|\right) \qquad G_{b.u.o1.e} = 14.234 \times 10^0 \quad [dB]$$

$$G_{b.u.o2} := -G_{b.u.o1} \qquad G_{b.u.o2} = 5.149 \times 10^0$$

$$G_{u.i1.b} := -2 \cdot g_{m.188}\left(\frac{1}{r_{a.188}} + \frac{1}{R_a}\right)^{-1} \qquad G_{u.i1.b} = -20.595 \times 10^0$$

$$G_{u.i2.b} := -G_{u.i1.b} \qquad G_{u.i2.b} = 20.595 \times 10^0$$

$$G_{u.i1.u.o1} := -g_{m.188}\left(\frac{1}{r_{a.188}} + \frac{1}{R_a}\right)^{-1} \qquad G_{u.i1.u.o1} = -10.298 \times 10^0$$

$$G_{u.i1.u.o2} := -G_{u.i1.u.o1} \qquad G_{u.i1.u.o2} = 10.298 \times 10^0$$

$$G_{u.i2.u.o2} := -g_{m.188}\left(\frac{1}{r_{a.188}} + \frac{1}{R_a}\right)^{-1} \qquad G_{u.i2.u.o2} = -10.298 \times 10^0$$

$$G_{u.i2.u.o1} := -G_{u.i2.u.o2} \qquad G_{u.i2.u.o1} = 10.298 \times 10^0$$

variation of R_{a1} will lead to the desired gain $G_{b.b}$, e.g a gain of 10 requires $R_{a1} = R_{a2} = 2k11535$

13.8.6 Specific impedances:

$$G_t := -\frac{1}{2} \cdot g_{m.188} \cdot \left(\frac{1}{r_{a.188}} + \frac{1}{R_a} \right)^{-1}$$

$$C_i := \left(1 + G_t \right) \cdot C_{g.a.1.188} + C_{g.c.1.188} + C_{stray.1.188}$$

$$Z_{i.b}(f) := \left(\frac{1}{2 \cdot R_g} + 2j \cdot \pi \cdot f \cdot 0.5 \cdot C_i \right)^{-1} \qquad\qquad Z_{i.u}(f) := \left(\frac{1}{R_g} + 2j \cdot \pi \cdot f \cdot C_i \right)^{-1}$$

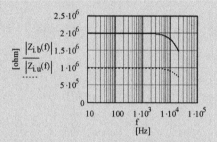

Figure 13.7 Input impedances

$$C_o := C_{a.c.1.188} \qquad R_{o.a.1} := \left(\frac{1}{r_{a.188}} + \frac{1}{R_{a1}} \right)^{-1} \qquad R_{o.a.2} := \left(\frac{1}{r_{a.188}} + \frac{1}{R_{a2}} \right)^{-1}$$

$$Z_{o.b}(f) := \left(\frac{1}{R_{o.a.1} + R_{o.a.2}} + 2j \cdot \pi \cdot f \cdot 0.5 \cdot C_o \right)^{-1} \qquad Z_{o.u}(f) := \left[\frac{1}{\left(\frac{1}{r_{a.188}} + \frac{1}{R_a} \right)^{-1}} + 2j \cdot \pi \cdot f \cdot C_o \right]^{-1}$$

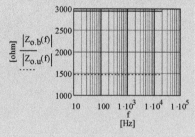

Figure 13.8 Output impedances

13.8.7 Gain stage frequency and phase response for the balanced input and balanced ouput case only:

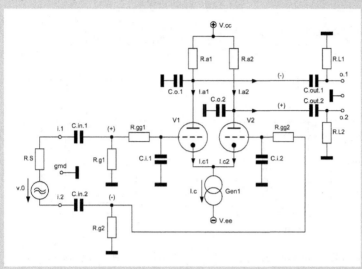

Figure 13.9 Balanced gain stage with all relevant components

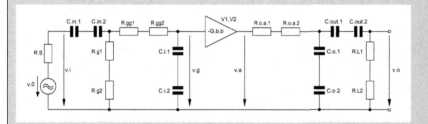

Figure 13.10 Simplified equivalent circuit of Figure 13.9

$$T_{i.b}(f) = \frac{v_g}{v0}$$

$$C_{i.1} := C_i \qquad C_{i.2} := C_i$$

$$Z1(f) := \frac{1}{2j \cdot \pi \cdot f \left(\dfrac{1}{C_{in.1}} + \dfrac{1}{C_{in.2}} \right)^{-1}}$$

$$Z2(f) := \frac{1}{2j \cdot \pi \cdot f \left(\dfrac{1}{C_{i.1}} + \dfrac{1}{C_{i.2}} \right)^{-1}}$$

$$T_{i.b}(f) := \frac{Z2(f) \cdot \left(\dfrac{1}{R_{g1} + R_{g2}} + \dfrac{1}{R_{gg1} + R_{gg2} + Z2(f)} \right)^{-1}}{\left(Z2(f) + R_{gg1} + R_{gg2} \right) \left[R_S + Z1(f) + \left(\dfrac{1}{R_{g1} + R_{g2}} + \dfrac{1}{R_{gg1} + R_{gg2} + Z2(f)} \right)^{-1} \right]}$$

$$T_{i.b.e}(f) := 20 \cdot \log\left(\left| T_{i.b}(f) \right| \right) \qquad\qquad \phi_{i.b}(f) := \operatorname{atan}\left(\frac{\operatorname{Im}\left(T_{i.b}(f) \right)}{\operatorname{Re}\left(T_{i.b}(f) \right)} \right)$$

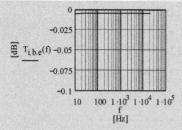

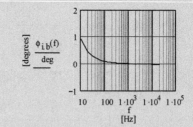

Figure 13.11 Transfer of input network   Figure 13.12 Phase of input network

$$T_{o.b}(f) = \frac{v_o}{v_a} \qquad\qquad R_{L.eff} := \left(\frac{1}{R_{L1}} + \frac{1}{R_{L2}} \right)^{-1} \qquad\qquad R_{a.eff} := \left(\frac{1}{R_a} + \frac{1}{R_{L.eff}} \right)^{-1}$$

$$G_{b.b.eff} := -g_{m.188} \cdot \left(\frac{1}{r_{a.188}} + \frac{1}{R_{a.eff}} \right)^{-1} \qquad\qquad G_{b.b.eff} = -10.234 \times 10^0$$

$$C_{o.1} := C_{a.c.1.188} \qquad\qquad C_{o.2} := C_{a.c.2.188}$$

$$Z3(f) := \frac{1}{2j \cdot \pi \cdot f \cdot \left(\dfrac{1}{C_{o.1}} + \dfrac{1}{C_{o.2}} \right)^{-1}} \qquad\qquad Z4(f) := \frac{1}{2j \cdot \pi \cdot f \cdot \left(\dfrac{1}{C_{out.1}} + \dfrac{1}{C_{out.2}} \right)^{-1}}$$

$$T_{o.b}(f) := \frac{\left(\dfrac{1}{Z3(f)} + \dfrac{1}{Z4(f) + R_{L1} + R_{L2}} \right)^{-1}}{R_{o.a.1} + R_{o.a.2} + \left(\dfrac{1}{Z3(f)} + \dfrac{1}{Z4(f) + R_{L1} + R_{L2}} \right)^{-1}} \cdot \frac{R_{L1} + R_{L2}}{R_{L1} + R_{L2} + Z4(f)}$$

$$T_{o.b.e}(f) := 20 \cdot \log\left(\left| T_{o.b}(f) \right| \right) \qquad\qquad \phi_{o.b}(f) := \operatorname{atan}\left(\frac{\operatorname{Im}\left(T_{o.b}(f) \right)}{\operatorname{Re}\left(T_{o.b}(f) \right)} \right)$$

➢   MCD Worksheet XIII   BAL calculations   Page 6

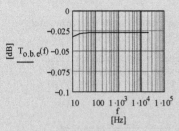

Figure 13.13 Transfer of output network

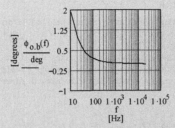

Figure 13.14 Phase of output network

$$G_{tot.b.b}(f) := T_{i.b}(f) \cdot T_{o.b}(f) \cdot G_{b.b.eff}$$

$$\phi_{G.b.b}(f) := -180 \, deg$$

$$G_{tot.b.b.e}(f) := 20 \cdot \log\left(\left|G_{tot.b.b}(f)\right|\right)$$

$$\phi_{tot.b.b}(f) := \phi_{i.b}(f) + \phi_{o.b}(f) + \phi_{G.b.b}(f)$$

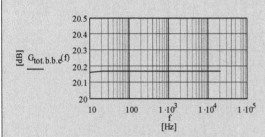

Figure 13.15 Frequency response of the whole gain stage

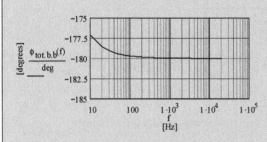

Figure 13.16 Phase response of the whole gain stage

$$G_{tot.b.b.e}(20Hz) = 20.168 \times 10^0 \quad [dB]$$

$$G_{tot.b.b.e}\left(10^3 Hz\right) = 20.17 \times 10^0 \quad [dB]$$

$$G_{tot.b.b.e}\left(20 \cdot 10^3 Hz\right) = 20.17 \times 10^0 \quad [dB]$$

$$\frac{\phi_{tot.b.b}(20Hz)}{deg} = -178.588 \times 10^0$$

$$\frac{\phi_{tot.b.b}\left(10^3 Hz\right)}{deg} = -179.974 \times 10^0$$

$$\frac{\phi_{tot.b.b}\left(20 \cdot 10^3 Hz\right)}{deg} = -180.049 \times 10^0$$

13.9   CMRR:

In a first example calculation it is assumed that the current generator Gen.1 is built-up by solid state active components, thus, the value of its internal resistance $r_{c.1}$ can be expected in the region of app. 1 - 2 MΩ.

A second example calculation uses a valve driven current generator with a much lower valued internal resistance of $r_{c.2}$ in the region of app. 150 kΩ. To enable the inclusion of this kind of solution the supply voltages have to be adapted: either very much higher V_{cc} or a rather high V_{ee}.

$$r_{c.1} := 1.5 \cdot 10^6 \, \Omega \qquad\qquad G_{com} = -\mu_{188} \cdot \frac{R_a}{r_{a.188} + R_a + \left(1 + \mu_{188}\right) \cdot 2 \cdot r_c}$$

$$r_{c.2} := 150 \cdot 10^3 \, \Omega$$

$R_{a.com}$ includes the frequency dependent effects of the whole output network:

$$R_{a.com}(f) := \left[\frac{1}{R_a} + \cfrac{1}{R_{L1} + R_{L2} + \cfrac{1}{2j \cdot \pi \cdot f \cdot \left(\cfrac{1}{C_{out.1}} + \cfrac{1}{C_{out.2}} \right)^{-1}}} \right]^{-1}$$

$$G_{com.eff.1}(f) := -\mu_{188} \cdot \frac{R_{a.com}(f)}{r_{a.188} + R_{a.com}(f) + \left(1 + \mu_{188}\right) \cdot 2 \cdot r_{c.1}}$$

$$G_{com.eff.2}(f) := -\mu_{188} \cdot \frac{R_{a.com}(f)}{r_{a.188} + R_{a.com}(f) + \left(1 + \mu_{188}\right) \cdot 2 \cdot r_{c.2}}$$

$$CMRR = \frac{G_{b.b}}{G_{com}} \qquad\qquad CMRR_1(f) := \frac{G_{tot.b.b}(f)}{G_{com.eff.1}(f)} \qquad\qquad CMRR_2(f) := \frac{G_{tot.b.b}(f)}{G_{com.eff.2}(f)}$$

$$CMRR_{1.e}(f) := 20 \cdot \log\left(CMRR_1(f)\right) \qquad\qquad CMRR_{2.e}(f) := 20 \cdot \log\left(CMRR_2(f)\right)$$

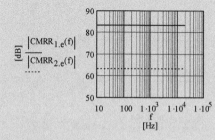

$$[dB] \quad \begin{array}{c} \left| CMRR_{1.e}(f) \right| \\ \hline \left| CMRR_{2.e}(f) \right| \\ \cdots\cdots \end{array}$$

Figure 13.17 CMRR of the balanced gain stage with two different types of current generators

➢ MCD Worksheet XIII BAL calculations Page 8

$$\left| CMRR_{1.e}\left(10^3 Hz\right)\right| = 83.123 \times 10^0 \quad [dB] \qquad \left| CMRR_{2.e}\left(10^3 Hz\right)\right| = 63.128 \times 10^0 \quad [dB]$$

Application of the rule of thumb (rot):

$$CMRR_{1.rot} := g_{m.188} \cdot r_{c.1} \qquad\qquad CMRR_{2.rot} := g_{m.188} \cdot r_{c.2}$$

$$CMRR_{1.rot.e} := 20 \cdot \log\left(CMRR_{1.rot}\right) \qquad CMRR_{2.rot.e} := 20 \cdot \log\left(CMRR_{2.rot}\right)$$

$$CMRR_{1.rot.e} = 80.424 \times 10^0 \quad [dB] \qquad CMRR_{2.rot.e} = 60.424 \times 10^0 \quad [dB]$$

Note: Any tiny unbalance of the two halves of a balanced gain stage will create additional CMRR reductions

Chapter 14 Feedback (FB)

14.1 On Feedback: a very short story by simple means

Usually, any system with input and output, whether it's technological or biological, reacts to signals that are arriving at the system's input sensors. The system's reaction shown at its output is based on these signals that are summed up with other system input signals generated by the system itself. To change the system output signal in a specific way a portion of the output signal can be fed back to the input and got summed up with all the other input signals, thus, influencing the input signals in a way that the whole system produces a corrected or controlled new output signal. This is a typical feedback controlled system.

In other words: no feedback means no controlled system inherent influence over the input / output signals, feedback means influence! Whereas technological systems in nearly 100% of the cases[1] react in a definite mathematical / physical manner on feedback - human beings or nations or any other group of people sometimes don't[2].

Two different types of feedback are used in electronic circuits: positive or negative feedback or both together. Positive means that a fraction of the output signal got summed up with the input signal without 180° phase change of the respective fraction of the output signal. Negative means that a 180° phase changed fraction of the output signal got summed up with the input signal.

Compared with an output signal without feedback positive feedback will lead to an increase of the output signal - sometimes up to a controlled or wild oscillation. Basically, negative feedback will lead to a decreased output signal. Very positive effects of negative feedback will occur as well, like eg. predictable performance and accuracy of a gain stage, decrease of its distortion, linearization of the frequency and phase response, compensation of temperature and component ageing effects, etc..

To precisely control gain the following paragraphs will deal with negative feedback only.

---

[1]   Are there exceptions? maybe Schrödingers cat problem? Or any other phenomena of quantum physics?

[2]   Worth reading: Stanislaw Lem's "Fiasco", ISBN 0-15-630630-1

14.2 Electronic model of negative feedback

We've seen in Chapters 1 and 2 that the same basic circuit around one valve will drastically change its transfer characteristic when inserting only one capacitance at the right place: C_c parallel to R_c. Besides a lot of other positive and negative effects this measure cuts off the feedback path via R_c and it increases the gain of the stage in the CCS+Cc case. How does this work? The following op-amp based Figure 14.1 explains the whole correlations between input and output of a feedback controlled system. It will lead to the basic feedback factor β and the basic equation for the gain of a feedback system. Absolutely equal mechanics apply to valve amp stages (see paragraph 14.3 and MCD worksheet XIV-2). I will discuss two different cases. The first case works without any output resistance R_o of the amplifier ($R_o = 0R$), the second case offers an output resistance $R_o > 0R$.

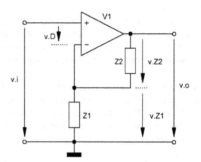

Figure 14.1 General feedback circuitry

14.2.1   Case 1: $R_o = 0R$

Assumed that the output resistance R_o of the amp V1 in Figure 14.1 is zero the derivation of the feedback equation looks as follows (G_o is the open loop gain of V1, G1 will be the gain of the Figure 14.1 gain stage after application of negative feedback from the output of V1 to its (-)-input):

$$v_o = G_o\, v_D$$
$$v_o = G_o\left(v_i - v_{Z1}\right) \tag{14.1}$$
$$v_o = G_o\left(v_i - v_o\,\frac{Z1}{Z1+Z2}\right)$$

$$\beta = \frac{Z1}{Z1+Z2} \tag{14.2}$$

$$\frac{v_o}{v_i} = G1 \tag{14.3}$$

$$G1 = \frac{G_o}{1+\beta\,G_o} \tag{14.4}$$

If G_o becomes nearly infinite the feedback controlled gain G1 of the stage will change to the well known equation for the gain of an op-amp gain stage:

$$G1 = 1 + \frac{Z2}{Z1} \tag{14.5}$$

14.2.2   Case 2: $R_o > 0R$

Assumed that the output resistance R_o of V1 is >0R, than, the above shown equations will change the following way (G_o is the open loop gain of V2, G2 will be the gain of the Figure 14.2 gain stage after application of negative feedback from the output of V1 to its (-)-input):

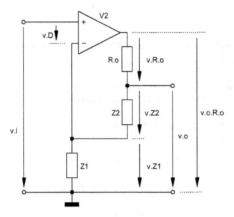

Figure 14.2 Feedback system with output resistance R_o bigger than zero

$$v_{o.R.o} = v_D \, G_o$$

$$v_o = v_D \, G_o - v_{R.o}$$

$$v_{R.o} = v_{o.R.o} \frac{R_o}{R_o + Z1 + Z2}$$

$$v_o = v_D \, G_o \left(1 - \frac{R_o}{R_o + Z1 + Z2} \right) \tag{14.6}$$

$$v_D = v_i - v_{Z1}$$

$$v_{Z1} = v_o \, \beta1$$

$$\beta1 = \frac{Z1}{Z1 + Z2}$$

$$\beta2 = \frac{R_o}{R_o + Z1 + Z2} \tag{14.7}$$

$$\Rightarrow \quad v_o = (v_i - v_o \beta 1) G_o (1 - \beta 2) \tag{14.8}$$

$$\frac{v_o}{v_i} = G2$$

$$\Rightarrow \quad G2 = \frac{G_o (1 - \beta 2)}{1 + G_o \beta 1 (1 - \beta 2)} \tag{14.9}$$

With $\beta 2 = 0$ and G_o = infinite the feedback controlled gain G2 of the stage becomes:

$$G2 = 1 + \frac{Z2}{Z1} \tag{14.10}$$

14.3   Feedback factor β1 for the CCS and CCS+Cc gain stage case of Chapters 1 & 2[3]

Gain of the CCS: $G1_u$ (see equation (1.2))
Gain of the CCS+Cc: $G1_b$ (see equation (2.2))

Thus, the feedback factor $\beta 1$ becomes[4]:

$$G1_u = \frac{G1_b}{1 + (1 + \mu) \dfrac{R_c}{r_a + R_a}} \tag{14.11}$$

$$\Rightarrow \quad (1 + \mu) \frac{R_c}{r_a + R_a} = \beta 1 \, G1_b \tag{14.12}$$

$$\Rightarrow \quad \beta 1 = -\left(\frac{1}{\mu} + 1 \right) \frac{R_c}{R_a} \tag{14.13}$$

Rule of thumb:

With $\mu \gg 1$ $\beta 1_{rot}$ becomes:

$$\beta 1_{rot} = -\frac{R_c}{R_a} \tag{14.14}$$

---

[3]   Details see MCD worksheet XIV-2 on the following pages
[4] R_c and R_a: see Chapter 2, Figure 2.1

14.4 Feedback with zero output resistance R_o:

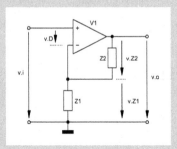

Figure 14.3 General feedback circuitry

$R_o = 0R$

$v_o = G_o \cdot v_D$

$v_o = G_o \cdot (v_i - v_{Z1})$

$v_o = G_o \cdot \left(v_i - v_o \cdot \dfrac{Z1}{Z1 + Z2}\right) = G_o \cdot v_i - G_o \cdot v_o \cdot \dfrac{Z1}{Z1 + Z2}$

$\dfrac{v_o}{v_i} = G1 = \dfrac{G_o}{1 + \beta1 \cdot G_o}$

$v_D = v_i - v_{Z1}$

$v_{Z1} = v_o \cdot \dfrac{Z1}{Z1 + Z2}$

$\beta1 = \dfrac{Z1}{Z1 + Z2}$

with $G_0 = \infty$ G1 becomes:

$G1 = \dfrac{1}{\dfrac{1}{G_o} + \beta1} = \dfrac{1}{\beta1} = \dfrac{Z1 + Z2}{Z1} = 1 + \dfrac{Z2}{Z1}$

14.5   Feedback with output resistance $R_o > 0 R$:

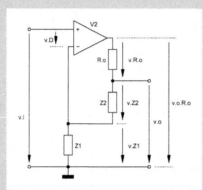

Figure 14.4 Feedback system with output
resistance bigger than zero

$$R_o = R$$

$$v_{o.R.o} = v_D \cdot G_o \qquad\qquad\qquad v_o = v_D \cdot G_o - v_{R.o}$$

$$v_{R.o} = v_{o.R.o} \frac{R_o}{R_o + Z1 + Z2} \qquad\qquad v_o = v_D \cdot G_o - v_{o.R.o} \frac{R_o}{R_o + Z1 + Z2}$$

$$v_D = v_i - v_{Z1} \qquad\qquad\qquad v_o = v_D \cdot G_o \left(1 - \frac{R_o}{R_o + Z1 + Z2}\right)$$

$$v_{Z1} = v_o \cdot \beta 1 \qquad\qquad\qquad \beta 2 = \frac{R_o}{R_o + Z1 + Z2}$$

$$v_o = \left(v_i - v_o \cdot \beta 1\right) \cdot G_o \cdot (1 - \beta 2) \qquad v_o \cdot \left(1 + G_o \cdot \beta 1 - G_o \cdot \beta 1 \cdot \beta 2\right) = v_i \cdot \left(G_o - G_o \cdot \beta 2\right)$$

$$\frac{v_o}{v_i} = G2 \qquad\qquad\qquad G2 = \frac{G_o(1 - \beta 2)}{1 + G_o \cdot \beta 1 \cdot (1 - \beta 2)}$$

with $\beta 2 = 0$ and $G_o = \infty$ G2 becomes:

$$G2 = \frac{1}{\dfrac{1}{G_o} + \beta 1} = \frac{1}{\beta 1} = \frac{Z1 + Z2}{Z1} = 1 + \frac{Z2}{Z1}$$

14.6   Example calculations:

14.6.1 $R_{o.1} = 0R$

$G_{o.1} := 2.074 \cdot 10^3$ $Z1 := 2.21 \cdot 10^3 \Omega$ $Z2 := 230 \cdot 10^3 \Omega$ $\beta1 := \dfrac{Z1}{Z1 + Z2}$

$G1 := \dfrac{G_{o.1}}{1 + \beta1 \cdot G_{o.1}}$ $G1 = 100.006$

$20 \cdot \log(G1) = 40.001$

14.6.2 $R_{o.2} > 0R$:
14.6.2.1   Variable $R_{o.2}$, fixed $G_{o.2}$:

$R_{o.2} := 10\Omega, 20\Omega .. 100000\,\Omega$ $G_{o.2} := G_{o.1}$ $\beta2(R_{o.2}) := \dfrac{R_{o.2}}{R_{o.2} + Z1 + Z2}$

$G2(R_{o.2}) := \dfrac{G_{o.2} \cdot (1 - \beta2(R_{o.2}))}{1 + \beta1 \cdot G_{o.2} \cdot (1 - \beta2(R_{o.2}))}$

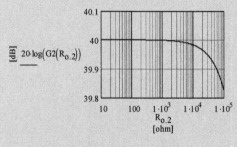

Figure 14.5 Total gain G2 vs. output resistance $R_{o.2}$

14.6.2.2   Fixed $R_{o.2}$, variable $G_{o.2}$:

$R_{o.2} := 1.5 \cdot 10^3 \Omega$ $G_{o.2} := 100, 200 .. 100000$

$G2(G_{o.2}) := \dfrac{G_{o.2} \cdot (1 - \beta2(R_{o.2}))}{1 + \beta1 \cdot G_{o.2} \cdot (1 - \beta2(R_{o.2}))}$

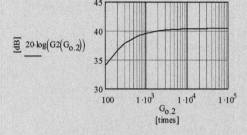

Figure 14.6 Total gain G2 vs. open loop gain $G_{o.2}$

14.7   Feedback factor $\beta 1$ of CCS (u) vs. CCS +Cc (b):

$$G1_u = -\mu \cdot \frac{R_a}{r_a + R_a + (1 + \mu) \cdot R_c} \qquad\qquad G1_b = -\mu \cdot \frac{R_a}{r_a + R_a}$$

$$\frac{G1_u}{G1_b} = \frac{\left[-\mu \cdot \dfrac{R_a}{r_a + R_a + (1 + \mu) \cdot R_c} \right]}{\left(-\mu \cdot \dfrac{R_a}{r_a + R_a} \right)}$$

$$\frac{G1_u}{G1_b} = \frac{r_a + R_a}{r_a + R_a + R_c + R_c \cdot \mu} \qquad\qquad \frac{G1_u}{G1_b} = \frac{1}{1 + (1 + \mu)\dfrac{R_c}{r_a + R_a}}$$

$$G1_u = \frac{G1_b}{\left[1 + (1 + \mu)\dfrac{R_c}{r_a + R_a} \right]} \qquad\qquad (1 + \mu)\frac{R_c}{r_a + R_a} = \beta \cdot G1_b$$

$$\beta 1 = \frac{(1 + \mu)\dfrac{R_c}{r_a + R_a}}{-\mu \cdot \dfrac{R_a}{r_a + R_a}} \qquad\qquad \beta 1 = -(1 + \mu) \cdot \frac{R_c}{\mu \cdot R_a} = -\left(\frac{1}{\mu} + 1 \right) \frac{R_c}{R_a}$$

$$G = \frac{G_o}{1 + \beta \cdot G_o} \qquad\qquad G1_u = \frac{G1_b}{1 + \beta 1 \cdot G1_b}$$

14.7.1 Example:

$$\mu := 100 \qquad\qquad R_c := 1.2 \cdot 10^3 \Omega \qquad\qquad R_a := 100 \cdot 10^3 \Omega \qquad\qquad r_a := 62.5 \cdot 10^3 \Omega$$

$$\beta 1 := -\left(\frac{1}{\mu} + 1 \right) \cdot \frac{R_c}{R_a} \qquad\qquad \beta 1 = -12.12 \times 10^{-3} \qquad \frac{R_c}{R_a} = 12 \times 10^{-3}$$

$$G1_b := -\mu \cdot \frac{R_a}{r_a + R_a} \qquad\qquad G1_b = -61.538$$

$$G1_u := \frac{G1_b}{1 + \beta 1 \cdot G1_b} \qquad\qquad G1_u = -35.249$$

14.7.2 Test:

$$G1_t := -\mu \cdot \frac{R_a}{r_a + R_a + (1 + \mu) \cdot R_c} \qquad\qquad G1_t = -35.249$$

Chapter 15 Design Example (EX)

Design goal: Development of a 3 stage RIAA-equalized and a 3 stage non-equalized pre-amp, each working alternately with overall and local feedback

15.1 Introduction

The aim of this rather complex example is the following one:

I want to check if there are differences that can be heard or otherwise detected between various feedback or non-feedback driven versions of two types of pre-amps -

- with equally biased valves in each version
- with no change of the DC environment of each valve
- with no change in overall gain at 1kHz (= 50 = +34dB) for the 4 differently designed RIAA equalized pre-amps
- with no change in overall gain at 1kHz (= 500 = +54dB) for the 4 differently designed pre-amps with a flat frequency response producing equalization

The pre-amp versions should look as follows:

> **A:** 4 different pre-amps à la Figures 15.1 and 15.4 with a **non-equalized** (flat) frequency response to be tested with test sequence 1 à la Figure 15.3:

> **B:** 4 different pre-amps à la Figures 15.1 and 15.4 with a **RIAA equalized** ($_r$ = RIAA equalized) frequency response to be tested with test sequence 2 à la Figure 15.6:

The 4 different sub-versions cc ... oo should look as follows:

- cc = overall feedback plus local feedback around each valve
- co = overall feedback plus no local feedback for V2 only
- oc = no overall feedback plus local feedback for V1 and V2
- oo = no overall feedback plus no local feedback for V1 and V2

> as CF1 in all versions V3‖V4 keep their current feedback via the cathode

> Version subscripts cc, co, oc, oo:

- 1st letter of the subscript indicates "overall"
- 2nd letter of the subscript indicates "local"
- c means: closed loop
- o means: open loop

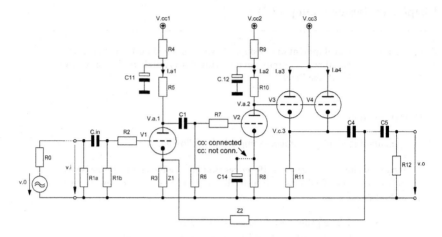

Figure 15.1 Basic pre-amp design with overall feedback - including
the alternative to change V2's gain and local feedback situation

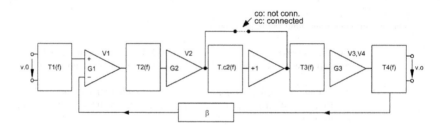

Figure 15.2 Simplified circuit diagram of Figure 15.1 including all calculation relevant
active and passive circuit blocks

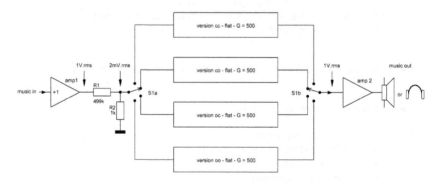

Figure 15.3 A-versions and test sequence 1

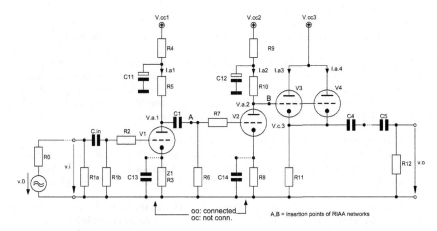

Figure 15.4 Basic pre-amp design without overall feedback - including
the alternative to change V1's and V2's gain and local feedback situation

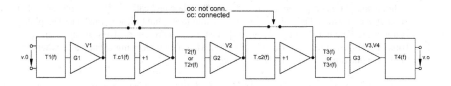

Figure 15.5 Simplified circuit diagram of Figure 15.4 including all calculation relevant
active and passive circuit blocks

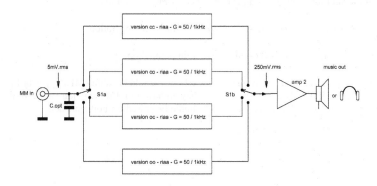

Figure 15.6 B-versions and test sequence 2

15.2 Description of the pre-amps

The purpose of this chapter is to demonstrate the usage of this reference book by going through the whole calculation course for the pre-amps (see MCD Worksheets XV-1 ... 3 on the following pages). The design goal will be to get pre-amps with a reasonable frequency and phase response in B_{20k} as well as - in the overall feed-back cases - an open loop gain G_o of approximately 1400 ... 1600, but not taking into account any feedback nor RIAA equalization. In the non-overall feed-back cases the overall gain should be 50 in the RIAA case and 500 in the flat frequency response case.

After inclusion of a suitable $Z2(f)$ forming resistor into the feedback path - or by adequately changing the plate resistors - the A-version pre-amps should produce $1V_{rms}$ output voltage with a $2mV_{rms}$ input signal at 1kHz. A deviation from a flat frequency response of max. -0.25 dB at the ends of the frequency band should be a result as well.

After inclusion of a suitable RIAA network[1] as $Z2(f)$ into the feedback path or as passive networks between V1 ... V3 the B-version pre-amps should produce $250V_{rms}$ output voltage with a $5mV_{rms}$ input signal at 1kHz. A deviation of max. ± 0.25 dB from the exact RIAA transfer should be the outcome.

The basic pre-amp consists of three active gain stages V1 ... V3,4 with respective gains G1 ... G3. The gains for V1 and V2 are set by - from version to version - different cathode and plate component arrangements. The plate and cathode DC currents as well as the plate and cathode DC voltages do not change throughout all versions! The DC plate voltages are kept constant by adequate split of the plate resistors. They always sum up to 100k. That's why the upper resistor needs an AC voltage blocking capacitance whereas the lower one defines the gain of that valve.

At the inputs and outputs of V1 ... V3,4 we find different stage separating passive networks. They serve for DC decoupling and valve bias setting as well as for frequency (also passive RIAA) and phase influencing purposes. AC voltage short-cutting cathode capacitances are also included into the calculation course (see Figures 15.2 & 15.5)

Some additional remarks on the components of the circuit diagrams:

- V1,2:   ECC83 - paired versions for stereo operation

- V3,4:   ECC83 - both systems in parallel operation

- non electrolytic capacitances C_{in}, C1 ... 3[2] could be Wima MKP types or Epcos B32520 ... B32529[3] (best case); they all have to fulfil the requirements of the rather high DC voltages. C2, C3 should be paired for both channels

---

[1]   for detailed calculations of RIAA networks see Chapters 8 & 9 of TSOS
[2]   for C2 = C_x, C3 = C_y see MCD Worksheet XV-1 on one of the following pages
[3]   for extremely low distortion types see Cyril Bateman's articles on "Understanding Capacitors" in Electronics World (EW) 12-1997 ... 08-1998 and "Capacitor Sounds" in EW 05 ... 11-2002

- the electrolytic capacitances[4] C4, C11, C12 could be low ESR 450V types

- C13, C14 could be 10V Panasonic FC

- C5 could be a 100V Panasonic FC

- all resistors should be metal types 1% E96 and should fulfil all kinds of specific power requirements; there is no need to take high price bulk foil resistors for R5, R10: the Signal-to-Noise-ratio improvement would be max. 0.2dB only

- for MM phono pre-amps of the shown types there is no need to include C_{in} into the circuitry, nevertheless I've done it to demonstrate its influence on frequency and phase response

- to make things a bit more difficult

  o I didn't take C4 and Z2 in a sequence and C5 direct connected to the cathodes of V3,4: the effect would be a simpler calculation approach and C4 would become a smaller size

  o I've included R2 and R7 to demonstrate the influence of these resistors - quite often used by designers to stop any kind of wild oscillation

  o I've also included a third wild oscillation preventing resistor R13 into the calculation course - it's not shown in Figures 15.1 & 15.4 between plate of V2 and grids of V3,4 - but we never know ...

---

[4] Test circuit for the selection of electrolytic capacitors by sound differences: see Figure 6.5 of TSOS
plus: worth to read the debate on electrolytic capacitors inside the audio chain: letters from Douglas Self and others in EW 04, 06, 10-1988

15.3   MathCad (MCD) worksheets for the Design Example - given on the following pages:

➤   MCD Worksheet XV-1

This worksheet offers the complete calculation course for the pre-amp type (with overall feedback) that serves for 4 different versions: two for the versions with a actively produced flat frequency response, two for the actively RIAA equalized versions.

➤   MCD Worksheet XV-2

This worksheet offers the complete calculation course for the pre-amp type (with local feedback only) that serves for 2 different versions that produce a flat frequency response.

➤   MCD Worksheet XV-3

This worksheet offers the complete calculation course for the pre-amp type (with local feedback only) that serves for 2 different versions that produce a passively RIAA equalized frequency response.

15.4   Example calculations for a 3-stage pre-amp with flat frequency response (versions $A_{cc}+A_{co}$) and a 3-stage MM phono pre-amp (versions $B_{cc}+B_{co}$)

15.4.1 Circuit diagram and components:

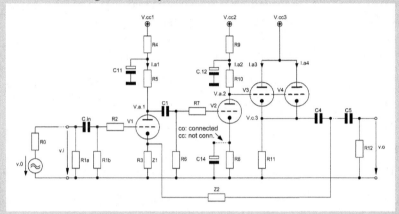

Figure 15.7 Basic pre-amp design with overall feedback (= Fig. 15.1)

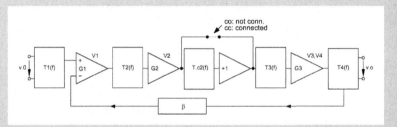

Figure 15.8 Simplified equivalent circuit of Figure 15.7 (= Fig. 15.2)

15.4.1.1   Valve constants and data sheet figures:

V1 ... 4: 1/2 ECC83-12AX7

$I_{a1} = I_{a2} = I_{a3} = I_{a4} = I_a$ $I_a := 1.7 \cdot 10^{-3} A$

$V_{cc1.cc} = V_{cc1.co} = V_{cc2.cc} = V_{cc2.co} = 420V$ $V_{cc3} = 420V$

$V_{a.1} = 250V$ $V_{a.2} = 250V$ $V_{a.3} = V_{cc3}$

$g_{m.v1} = g_{m.v2} = g_{m.v3} = g_{m.v4} = g_m$ $g_m := 1.9 \cdot 10^{-3} S$

$\mu_{v1} = \mu_{v2} = \mu_{v3} = \mu_{v4} = \mu$ $\mu := 101$

$r_{a.v1} = r_{a.v2} = r_{a.v3} = r_{a.v4} = r_a$ $r_a := 53 \cdot 10^3 \Omega$

$C_{g.a.83} := 1.6 \cdot 10^{-12} F$ $C_{a.c.83} := 0.33 \cdot 10^{-12} F$ $C_{g.c.83} := 1.65 \cdot 10^{-12} F$

➢ **MCD Worksheet XV-1**   overall feedback versions   page 2

15.4.1.2   Component figures:

$R0 := 1 \cdot 10^3 \Omega$ $R1a := 52.3 \cdot 10^3 \Omega$ $R1b := 475 \cdot 10^3 \Omega$

                                               $R2 := 100 \Omega$ $R3 := 1 \cdot 10^3 \Omega$

$R6 := 1000 \cdot 10^3 \Omega$ $R7 := 100 \Omega$ $R8 := 1 \cdot 10^3 \Omega$

$R11 := 75 \cdot 10^3 \Omega$ $R12 := 47.5 \cdot 10^3 \Omega$ $R13 := 100 \Omega$

$R5_{cc} := 100 \cdot 10^3 \Omega$ $R4_{cc} := 100 \cdot 10^3 \Omega - R5_{cc}$ $R4_{cc} = 0 \times 10^0 \Omega$

$R5_{co} := R5_{cc}$ $R4_{co} := 100 \cdot 10^3 \Omega - R5_{co}$ $R4_{co} = 0 \times 10^0 \Omega$

$R10_{cc} := 100 \cdot 10^3 \Omega$ $R9_{cc} := 100 \cdot 10^3 \Omega - R10_{cc}$ $R9_{cc} = 0 \times 10^0 \Omega$

$R10_{co} := 34 \cdot 10^3 \Omega$ $R9_{co} := 100 \cdot 10^3 \Omega - R10_{co}$ $R9_{co} = 66 \times 10^3 \Omega$

$C_{in} := 2.2 \cdot 10^{-6} F$ $C1 := 470 \cdot 10^{-9} F$ $C4 := 22 \cdot 10^{-6} F$

$C5 := 22 \cdot 10^{-6} F$ $C11 = 10 \cdot 10^{-6} F$ $C12 = C11$

$C_{stray.1} := 10 \cdot 10^{-12} F$ $C_{stray.2} := 5 \cdot 10^{-12} F$ $C_{stray.3} := 4 \cdot 10^{-12} F$

15.4.2 Gain calculations (Z1, Z2 purely resistive)

$Z1 := R3$ $Z2 := 0.8 \cdot 10^6 \Omega$

$$G1_{cc} := -\mu \cdot \frac{\left(\dfrac{1}{R5_{cc}} + \dfrac{1}{R6}\right)^{-1}}{r_a + \left(\dfrac{1}{R5_{cc}} + \dfrac{1}{R6}\right)^{-1} + (1+\mu) \cdot R3}$$ $G1_{cc} = -37.338 \times 10^0$

$G1_{co} := G1_{cc}$ $G1_{co} = -37.338 \times 10^0$

$$G2_{cc} := -\mu \cdot \frac{R10_{cc}}{r_a + R10_{cc} + (1+\mu) \cdot R8}$$ $G2_{cc} = -39.608 \times 10^0$

$$G2_{co} := -\mu \cdot \frac{R10_{co}}{r_a + R10_{co}}$$ $G2_{co} = -39.471 \times 10^0$

$$r_{a.v3.4} := \frac{r_a}{2}$$

$$G3 := \mu \cdot \frac{R11}{r_{a.v3.4} + (1+\mu) \cdot R11}$$ $G3 = 986.778 \times 10^{-3}$

$A_{cc} := G1_{cc} \cdot G2_{cc} \cdot G3$ $A_{cc} = 1.459 \times 10^3$

$A_{co} := G1_{co} \cdot G2_{co} \cdot G3$ $A_{co} = 1.454 \times 10^3$

$$\beta(Z2) := \frac{Z1}{Z1+Z2} \qquad\qquad \beta(Z2) = 1.248 \times 10^{-3}$$

$$A_{cc.tot}(Z2) := \frac{A_{cc}}{1+\beta(Z2)\cdot A_{cc}} \qquad\qquad A_{cc.tot}(Z2) = 517.148 \times 10^0$$

$$A_{co.tot}(Z2) := \frac{A_{co}}{1+\beta(Z2)\cdot A_{co}} \qquad\qquad A_{co.tot}(Z2) = 516.514 \times 10^0$$

$$Z2_v := 100\cdot10^3\Omega, 110\cdot10^3\Omega .. 10^6\Omega$$

$$A_{cc.tot.e}(Z2_v) := 20\cdot\log(A_{cc.tot}(Z2_v)) \qquad A_{co.tot.e}(Z2_v) := 20\cdot\log(A_{co.tot}(Z2_v))$$

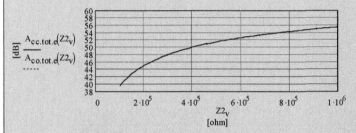

$$\begin{array}{c} [dB] \\ \dfrac{A_{cc.tot.e}(Z2_v)}{A_{co.tot.e}(Z2_v)} \\ ----- \end{array}$$

Figure 15.9 Plots of gains vs. Z2 of both pre-amps

15.4.3 Network calculations

$$f := 10Hz, 20Hz .. 20000\,Hz$$

$$f_{opt} := 0.2Hz$$

15.4.3.1   T1:

$$C_{i\,1.cc.tot} := \left(1+|G1_{cc}|\right)\cdot C_{g.a.83} + C_{g.c.83} + C_{stray.1}$$

$$Z1_1(f) := \frac{1}{2j\cdot\pi\cdot f\cdot C_{in}}$$

$$C_{i\,1.cc.tot} = 72.991 \times 10^{-12}F$$

$$Z2_{1.cc}(f) := \frac{1}{2j\cdot\pi\cdot f\cdot C_{i\,1.cc.tot}}$$

$$C_{i\,1.co.tot} := C_{i\,1.cc.tot}$$

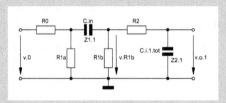

Figure 15.10 T1(f) network

➢   MCD Worksheet XV-1   overall feedback versions   page 4

$$T1(f) = \frac{v_{o.1}(f)}{v_0(f)} \qquad M_{cc}(f) := \frac{Z2_{1.cc}(f)}{R2 + Z2_{1.cc}(f)} \qquad \left(\frac{1}{R1a} + \frac{1}{R1b}\right)^{-1} = 47.113 \times 10^3 \, \Omega$$

$$N_{cc}(f) := \frac{\left(\dfrac{1}{R1b} + \dfrac{1}{R2 + Z2_{1.cc}(f)}\right)^{-1}}{Z1_1(f) + \left(\dfrac{1}{R1b} + \dfrac{1}{R2 + Z2_{1.cc}(f)}\right)^{-1}}$$

$$O_{cc}(f) := \frac{\left[\dfrac{1}{R1a} + \dfrac{1}{\left[Z1_1(f) + \left(\dfrac{1}{R1b} + \dfrac{1}{R2 + Z2_{1.cc}(f)}\right)^{-1}\right]}\right]^{-1}}{R0 + \left[\dfrac{1}{R1a} + \dfrac{1}{\left[Z1_1(f) + \left(\dfrac{1}{R1b} + \dfrac{1}{R2 + Z2_{1.cc}(f)}\right)^{-1}\right]}\right]^{-1}}$$

$$T1_{cc}(f) := M_{cc}(f) \cdot N_{cc}(f) \cdot O_{cc}(f)$$

$$T1_{co}(f) := T1_{cc}(f)$$

$$T1_{cc.e}(f) := 20 \cdot \log\left(\left|T1_{cc}(f)\right|\right) \qquad\qquad \phi 1_{cc}(f) := \operatorname{atan}\left(\frac{\operatorname{Im}\left(T1_{cc}(f)\right)}{\operatorname{Re}\left(T1_{cc}(f)\right)}\right)$$

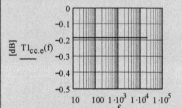

Figure 15.11 Transfers of T1(f)

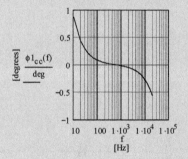

Figure 15.12 Phases of T1(f)

➢ MCD Worksheet XV-1 overall feedback versions page 5

15.4.3.2 T2:

$$R_{o.a.1.cc} := \left[\frac{1}{R5_{cc}} + \frac{1}{r_a + (1+\mu)R3} \right]^{-1}$$

$$R_{o.a.1.cc} = 60.784 \times 10^3 \, \Omega$$

$$R_{o.a.1.co} := \left[\frac{1}{R5_{co}} + \frac{1}{r_a + (1+\mu)R3} \right]^{-1}$$

$$R_{o.a.1.co} = 60.784 \times 10^3 \, \Omega$$

$$C_{o.1.cc.tot} := C_{g.a.83} + C_{a.c.83}$$

$$C_{i.2.cc.tot} := \left(1 + \left| G2_{cc} \right| \right) \cdot C_{g.a.83} + C_{stray.2} + C_{g.c.83}$$

$$C_{o.1.cc.tot} = 1.93 \times 10^{-12} \, F$$

$$C_{i.2.cc.tot} = 71.623 \times 10^{-12} \, F$$

$$C_{o.1.co.tot} := C_{g.a.83} + C_{a.c.83}$$

$$C_{i.2.co.tot} := \left(1 + \left| G2_{co} \right| \right) \cdot C_{g.a.83} + C_{stray.2} + C_{g.c.83}$$

$$C_{o.1.co.tot} = 1.93 \times 10^{-12} \, F$$

$$C_{i.2.co.tot} = 71.404 \times 10^{-12} \, F$$

$$Z1_{2.cc}(f) := \frac{1}{2j \cdot \pi \cdot f \cdot C_{o.1.cc.tot}}$$

$$Z2_2(f) := \frac{1}{2j \cdot \pi \cdot f \cdot C1}$$

$$Z3_{2.cc}(f) := \frac{1}{2j \cdot \pi \cdot f \cdot C_{i.2.cc.tot}}$$

$$Z1_{2.co}(f) := \frac{1}{2j \cdot \pi \cdot f \cdot C_{o.1.co.tot}}$$

$$Z3_{2.co}(f) := \frac{1}{2j \cdot \pi \cdot f \cdot C_{i.2.co.tot}}$$

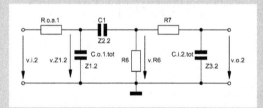

Figure 15.13 T2(f) network

$$v_{Z1.2}(f) = v_{i.2}(f) \cdot \frac{\left[\dfrac{1}{Z1_2(f)} + \dfrac{1}{Z2_2(f) + \left(\dfrac{1}{R6} + \dfrac{1}{Z3_2(f)} \right)^{-1}} \right]^{-1}}{R_{o.a.1} + \left[\dfrac{1}{Z1_2(f)} + \dfrac{1}{Z2_2(f) + \left(\dfrac{1}{R6} + \dfrac{1}{Z3_2(f)} \right)^{-1}} \right]^{-1}}$$

> MCD Worksheet XV-1   overall feedback versions   page 6

$$v_{R6}(f) = v_{Z.1.2}(f) \cdot \frac{\left(\dfrac{1}{R6} + \dfrac{1}{R7 + Z3_2(f)}\right)^{-1}}{Z2_2(f) + \left(\dfrac{1}{R6} + \dfrac{1}{R7 + Z3_2(f)}\right)^{-1}}$$

$$v_{o.2}(f) = v_{R6}(f) \cdot \frac{Z3_2(f)}{R7 + Z3_2(f)} \qquad\qquad\qquad T2(f) = \frac{v_{o.2}(f)}{v_{1.2}(f)}$$

$$T2_{1.cc}(f) := \frac{\left[\dfrac{1}{Z1_{2.cc}(f)} + \dfrac{1}{Z2_2(f) + \left(\dfrac{1}{R6} + \dfrac{1}{Z3_{2.cc}(f)}\right)^{-1}}\right]^{-1}}{R_{o.a.1.cc} + \left[\dfrac{1}{Z1_{2.cc}(f)} + \dfrac{1}{Z2_2(f) + \left(\dfrac{1}{R6} + \dfrac{1}{Z3_{2.cc}(f)}\right)^{-1}}\right]^{-1}}$$

$$T2_{1.co}(f) := \frac{\left[\dfrac{1}{Z1_{2.co}(f)} + \dfrac{1}{Z2_2(f) + \left(\dfrac{1}{R6} + \dfrac{1}{Z3_{2.co}(f)}\right)^{-1}}\right]^{-1}}{R_{o.a.1.co} + \left[\dfrac{1}{Z1_{2.co}(f)} + \dfrac{1}{Z2_2(f) + \left(\dfrac{1}{R6} + \dfrac{1}{Z3_{2.co}(f)}\right)^{-1}}\right]^{-1}}$$

$$T2_{2.cc}(f) := \frac{\left(\dfrac{1}{R6} + \dfrac{1}{R7 + Z3_{2.cc}(f)}\right)^{-1}}{Z2_2(f) + \left(\dfrac{1}{R6} + \dfrac{1}{R7 + Z3_{2.cc}(f)}\right)^{-1}} \cdot \frac{Z3_{2.cc}(f)}{R7 + Z3_{2.cc}(f)}$$

$$T2_{2.co}(f) := \frac{\left(\dfrac{1}{R6} + \dfrac{1}{R7 + Z3_{2.co}(f)}\right)^{-1}}{Z2_2(f) + \left(\dfrac{1}{R6} + \dfrac{1}{R7 + Z3_{2.co}(f)}\right)^{-1}} \cdot \frac{Z3_{2.co}(f)}{R7 + Z3_{2.co}(f)}$$

➢   MCD Worksheet XV-1   overall feedback versions   page 7

15.4.3.3   Tc2:

$Tc2_{cc}(f) := 1$

$r_{c.2.co} := \dfrac{R10_{co} + r_a}{\mu + 1}$

$R_{o.c2.co} := \left(\dfrac{1}{r_{c.2.co}} + \dfrac{1}{R8} \right)^{-1}$

$C14 := \dfrac{1}{2 \cdot \pi \cdot f_{opt} \cdot R_{o.c2.co}}$

$C14 = 1.729 \times 10^{-3}\,F$

$Tc2_{co}(f) := \dfrac{R_{o.c2.co}}{R_{o.c2.co} + \left(2j \cdot \pi \cdot f \cdot C14 \right)^{-1}}$

$T2_{cc}(f) := T2_{1.cc}(f) \cdot T2_{2.cc}(f) \cdot Tc2_{cc}(f)$

$T2_{cc.e}(f) := 20 \cdot \log\left(\left| T2_{cc}(f) \right| \right)$

$\phi 2_{cc}(f) := \operatorname{atan}\left(\dfrac{\operatorname{Im}\left(T2_{cc}(f) \right)}{\operatorname{Re}\left(T2_{cc}(f) \right)} \right)$

$T2_{co}(f) := T2_{1.co}(f) \cdot T2_{2.co}(f) \cdot Tc2_{co}(f)$

$T2_{co.e}(f) := 20 \cdot \log\left(\left| T2_{co}(f) \right| \right)$

$\phi 2_{co}(f) := \operatorname{atan}\left(\dfrac{\operatorname{Im}\left(T2_{co}(f) \right)}{\operatorname{Re}\left(T2_{co}(f) \right)} \right)$

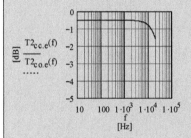

Figure 15.14 Transfers of T2(f)

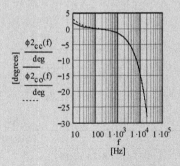

Figure 15.15 Phases of T2(f)

15.4.3.4   T3:

$R_{o.a.2.cc} := \left[\dfrac{1}{R10_{cc}} + \dfrac{1}{r_a + (1 + \mu)R8} \right]^{-1}$

$R_{o.a.2.cc} = 60.784 \times 10^3\,\Omega$

$R_{o.a.2.co} := \left[\dfrac{1}{R10_{co}} + \dfrac{1}{r_a + (1 + \mu)R8} \right]^{-1}$

$R_{o.a.2.co} = 27.884 \times 10^3\,\Omega$

$C_{i.3.tot} := C_{g.a.83} + C_{stray.3} + (1 - |G3|) \cdot C_{g.c.83}$

$C_{i.3.tot} = 5.622 \times 10^{-12}\,F$

$C_{o.2.tot} := C_{a.c.83} + C_{g.a.83}$

$C_{o.2.tot} = 1.93 \times 10^{-12}\,F$

$C_{i.3.tot} := C_{g.a.83} + C_{stray.3} + (1 - |G3|) \cdot C_{g.c.83}$

$C_{i.3.tot} = 5.622 \times 10^{-12}\,F$

➢   MCD Worksheet XV-1   overall feedback versions   page 8

$$C_{o.2.co.tot} := C_{a.c.83} + C_{g.a.83}$$

$$C_{o.2.co.tot} = 1.93 \times 10^{-12} F$$

$$Z1_3(f) := \frac{1}{2j \cdot \pi \cdot f \cdot C_{o.2.tot}}$$

$$Z2_3(f) := \frac{1}{2j \cdot \pi \cdot f \cdot C_{i.3.tot}}$$

$$Z1_{3.co}(f) := \frac{1}{2j \cdot \pi \cdot f \cdot C_{o.2.co.tot}}$$

$$Z2_3(f) := \frac{1}{2j \cdot \pi \cdot f \cdot C_{i.3.tot}}$$

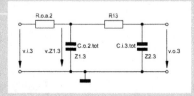

Figure 15.16 T3(f) network

$$v_{o.3}(f) = v_{Z.1.3}(f) \cdot \frac{Z2_3(f)}{Z2_3(f) + R13}$$

$$v_{Z1.3}(f) = v_{i.3}(f) \cdot \frac{\left(\dfrac{1}{Z1_3(f)} + \dfrac{1}{R13 + Z2_3(f)} \right)^{-1}}{R_{o.a.2} + \left(\dfrac{1}{Z1_3(f)} + \dfrac{1}{R13 + Z2_3(f)} \right)^{-1}}$$

$$T3(f) = \frac{v_{o.3}(f)}{v_{i.3}(f)}$$

$$T3_{cc}(f) := \frac{\left(\dfrac{1}{Z1_3(f)} + \dfrac{1}{R13 + Z2_3(f)} \right)^{-1}}{R_{o.a.2.cc} + \left(\dfrac{1}{Z1_3(f)} + \dfrac{1}{R13 + Z2_3(f)} \right)^{-1}} \cdot \frac{Z2_3(f)}{Z2_3(f) + R13}$$

$$T3_{co}(f) := \frac{\left(\dfrac{1}{Z1_{3.co}(f)} + \dfrac{1}{R13 + Z2_3(f)} \right)^{-1}}{R_{o.a.2.co} + \left(\dfrac{1}{Z1_{3.co}(f)} + \dfrac{1}{R13 + Z2_3(f)} \right)^{-1}} \cdot \frac{Z2_3(f)}{Z2_3(f) + R13}$$

$$T3_{cc.e}(f) := 20 \cdot \log\left(\left| T3_{cc}(f) \right| \right)$$

$$\phi 3_{cc}(f) := atan\left(\frac{Im\left(T3_{cc}(f) \right)}{Re\left(T3_{cc}(f) \right)} \right)$$

$$T3_{co.e}(f) := 20 \cdot \log\left(\left| T3_{co}(f) \right| \right)$$

$$\phi 3_{co}(f) := atan\left(\frac{Im\left(T3_{co}(f) \right)}{Re\left(T3_{co}(f) \right)} \right)$$

➤ MCD Worksheet XV-1   overall feedback versions   page 9

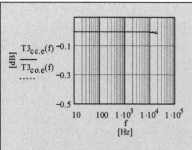

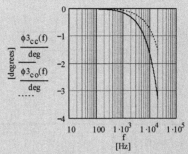

Figure 15.17 Transfers of T3(f)

Figure 15.18 Phases of T3(f)

15.4.3.5   T4:

$$r_{c.3} := \frac{r_{a.v3.4}}{\mu + 1}$$

$$r_{c.3} = 259.804 \times 10^0 \, \Omega$$

$$R_{o.c.3} := \left(\frac{1}{r_{c.3}} + \frac{1}{R11} \right)^{-1}$$

$$R_{o.c.3} = 258.907 \times 10^0 \, \Omega$$

$$r_{c.1} := \frac{r_a}{\mu + 1}$$

$$r_{c.1} = 519.608 \times 10^0 \, \Omega$$

$$R_{ff} := Z2 + \left(\frac{1}{R3} + \frac{1}{r_{c.1}} \right)^{-1}$$

$$R_{ff} = 800.342 \times 10^3 \, \Omega$$

$$Z1_4(f) := \frac{1}{2j \cdot \pi \cdot f \cdot C4}$$

$$Z2_4(f) := \frac{1}{2j \cdot \pi \cdot f \cdot C5}$$

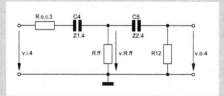

Figure 15.19 T4(f) network

$$v_{o.4}(f) = v_{R.ff} \frac{R12}{R12 + Z2_4(f)}$$

➤   MCD Worksheet XV-1   overall feedback versions   page 10

$$v_{R.ff}(f) = v_{i.4}(f) \cdot \frac{\left(\dfrac{1}{R_{ff}} + \dfrac{1}{R12 + Z2_4(f)}\right)^{-1}}{R_{o.c.3} + Z1_4(f) + \left(\dfrac{1}{R_{ff}} + \dfrac{1}{R12 + Z2_4(f)}\right)^{-1}} \qquad T4(f) = \frac{v_{o.4}(f)}{v_{i.4}(f)}$$

$$T4(f) := \frac{\left(\dfrac{1}{R_{ff}} + \dfrac{1}{R12 + Z2_4(f)}\right)^{-1}}{R_{o.c.3} + Z1_4(f) + \left(\dfrac{1}{R_{ff}} + \dfrac{1}{R12 + Z2_4(f)}\right)^{-1}} \cdot \frac{R12}{R12 + Z2_4(f)}$$

$$T4_e(f) := 20 \cdot \log\left(\left|T4(f)\right|\right) \qquad\qquad \phi4(f) := \operatorname{atan}\left(\frac{\operatorname{Im}(T4(f))}{\operatorname{Re}(T4(f))}\right)$$

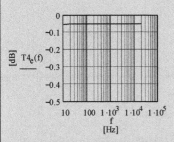

Figure 15.20 Transfers of T4(f)

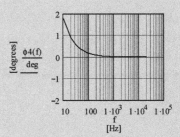

Figure 15.21 Phases of T4(f)

15.4.4 Total gains A$_{xy.tot}$(f) and phases $\phi_{xy.tot}$(f) for the "flat" pre-amps:

$$Z1(f) := 10^3 \Omega \qquad\qquad Z2(f) := 0.8 \cdot 10^6 \Omega$$

$$A_{cc}(f) := T1_{cc}(f) \cdot T2_{cc}(f) \cdot T3_{cc}(f) \cdot T4(f) \cdot A_{cc} \qquad\qquad \beta(f) := \frac{Z1(f)}{Z1(f) + Z2(f)}$$

$$A_{co}(f) := T1_{co}(f) \cdot T2_{co}(f) \cdot T3_{co}(f) \cdot T4(f) \cdot A_{co}$$

$$A_{cc.tot}(f) := \frac{A_{cc}(f)}{1 + \beta(f) \cdot A_{cc}(f)} \qquad\qquad A_{cc.tot.e}(f) := 20 \cdot \log\left(\left|A_{cc.tot}(f)\right|\right)$$

$$A_{co.tot}(f) := \frac{A_{co}(f)}{1 + \beta(f) \cdot A_{co}(f)} \qquad\qquad A_{co.tot.e}(f) := 20 \cdot \log\left(\left|A_{co.tot}(f)\right|\right)$$

> MCD Worksheet XV-1   overall feedback versions   page 11

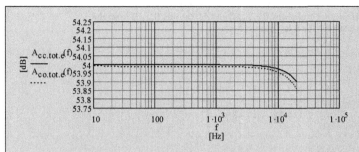

Figure 15.22 Frequency responses and gains of both "flat" pre-amps

$$\phi_{A.cc.tot}(f) := atan\left(\frac{Im\left(A_{cc.tot}(f)\right)}{Re\left(A_{cc.tot}(f)\right)}\right) \qquad \phi_{A.co.tot}(f) := atan\left(\frac{Im\left(A_{co.tot}(f)\right)}{Re\left(A_{co.tot}(f)\right)}\right)$$

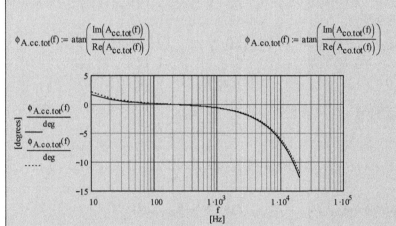

Figure 15.23 Phase responses of both "flat" pre-amps

$$A_{cc.tot.e}\left(10^3 Hz\right) = 54 \times 10^0 \qquad \frac{\phi_{A.cc.tot}(20Hz)}{deg} = 826.821 \times 10^{-3}$$

$$A_{co.tot.e}\left(10^3 Hz\right) = 53.989 \times 10^0 \qquad \frac{\phi_{A.co.tot}(20Hz)}{deg} = 1.044 \times 10^0$$

$$A_{cc.tot.e}\left(20 \cdot 10^3 Hz\right) = 53.903 \times 10^0 \qquad \frac{\phi_{A.cc.tot}\left(20 \cdot 10^3 Hz\right)}{deg} = -12.781 \times 10^0$$

$$A_{co.tot.e}\left(20 \cdot 10^3 Hz\right) = 53.861 \times 10^0 \qquad \frac{\phi_{A.co.tot}\left(20 \cdot 10^3 Hz\right)}{deg} = -12.055 \times 10^0$$

15.4.5 RIAA equalization, total gains $B_{xy.r.tot}(f)$ and deviations $D_{xy.r}(f)$ from the exact RIAA transfer $R_0(f)$ for both pre-amps:

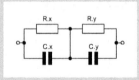

Figure 15.24 RIAA network = Z2(f) for feedback
equalization purposes

$R_x := 39.2 \cdot 10^3 \Omega$ $C_x := 2 \cdot 10^{-9} F$

$R_y := 750 \cdot 10^3 \Omega$ $C_y := 6.8 \cdot 10^{-9} F$

$Z1(f) := 10^3 \Omega$ $Z2_r(f) := \left(\dfrac{1}{R_x} + 2j \cdot \pi \cdot f C_x\right)^{-1} + \left(\dfrac{1}{R_y} + 2j \cdot \pi \cdot f C_y\right)^{-1}$

$\beta_r(f) := \dfrac{Z1(f)}{Z1(f) + Z2_r(f)}$

$B_{cc.r.tot}(f) := \dfrac{A_{cc}(f)}{1 + \beta_r(f) \cdot A_{cc}(f)}$

$B_{co.r.tot}(f) := \dfrac{A_{co}(f)}{1 + \beta_r(f) \cdot A_{co}(f)}$

$B_{cc.r.tot.e}(f) := 20 \cdot \log\left(\left|B_{cc.r.tot}(f)\right|\right)$ $B_{co.r.tot.e}(f) := 20 \cdot \log\left(\left|B_{co.r.tot}(f)\right|\right)$

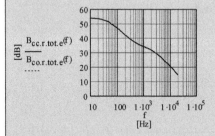

$\dfrac{B_{cc.r.tot.e}(f)}{\underset{\text{-----}}{B_{co.r.tot.e}(f)}}$ [dB]

Figure 15.25 RIAA
transfer of both pre-amps

$B_{cc.r.tot.e}\left(10^3 Hz\right) = 33.963 \times 10^0$

$B_{co.r.tot.e}\left(10^3 Hz\right) = 33.963 \times 10^0$

➢   MCD Worksheet XV-1   overall feedback versions   page 13

$B_{cc.r.tot.dif.e}(f) := B_{cc.r.tot.e}(f) - B_{cc.r.tot.e}\left(10^3 Hz\right)$

$B_{co.r.tot.dif.e}(f) := B_{co.r.tot.e}(f) - B_{co.r.tot.e}\left(10^3 Hz\right)$

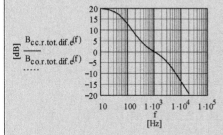

Figure 15.26 RIAA transfer of both pre-amps - ref. 0dB/1kHz

$$R_{1000} := \frac{\sqrt{1 + \left(2\cdot\pi\cdot 10^3 Hz\, 318\cdot 10^{-6} s\right)^2}}{\sqrt{1 + \left(2\cdot\pi\cdot 10^3 Hz\, 3180\cdot 10^{-6} s\right)^2}\,\sqrt{1 + \left(2\cdot\pi\cdot 10^3 Hz\, 75\cdot 10^{-6} s\right)^2}}$$

$$R_0(f) := R_{1000}^{-1}\cdot\frac{\sqrt{1 + \left(2\cdot\pi\cdot f\, 318\cdot 10^{-6} s\right)^2}}{\sqrt{1 + \left(2\cdot\pi\cdot f\, 3180\cdot 10^{-6} s\right)^2}\,\sqrt{1 + \left(2\cdot\pi\cdot f\, 75\cdot 10^{-6} s\right)^2}}$$

$D_{cc.r}(f) := 20\cdot\log\left(R_0(f)\right) - B_{cc.r.tot.dif.e}(f)$ $D_{co.r}(f) := 20\cdot\log\left(R_0(f)\right) - B_{co.r.tot.dif.e}(f)$

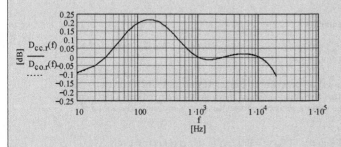

Figure 15.27 Deviations $D_{xy.r}(f)$ from the exact RIAA transfer $R_0(f)$ for both pre-amps

➤   MCD Worksheet XV-1   overall feedback versions   page 14

15.4.6 Input impedance $Z_{in.A.B.xy}(f)$ for the 4 versions

$$Z_{in.A.B.xy}(f) := \left[\frac{1}{R1a} + \cfrac{1}{\left[Z1_1(f) + \left(\frac{1}{R1b} + \frac{1}{R2 + Z2_{1.cc}(f)} \right)^{-1} \right]} \right]^{-1}$$

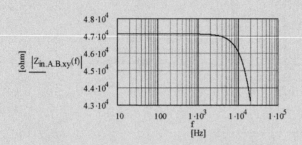

Figure 15.28 Input impedance for the 4 different versions

$$\left| Z_{in.A.B.xy}\left(10^3 Hz\right) \right| = 47.102 \times 10^3 \, \Omega \qquad \left| Z_{in.A.B.xy}\left(20 \cdot 10^3 Hz\right) \right| = 43.233 \times 10^3 \, \Omega$$

15.4.7   Final remarks:

15.4.7.1   Varying f_{opt} will lead to drastic changes in the low frequency range of the $D_{co.r}(f)$ plot - shown in Figure 15.27

15.4.7.2   To get lowest deviation from the exact RIAA transfer only R_x and R_y need further trimming - assumed that C_x and C_y got fixed values that were calculated with the respective formulae given in my book "The Sound of Silence"

15.4.7.3   A +/- 50% change of C_{in} will also lead to tiny changes in the low frequeny range of $D_{xy.r}(f)$ plot - shown in Figure 15.27

15.5 Example calculations for a 3-stage pre-amp ($A_{oc}+A_{oo}$ versions) with flat frequency response in B_{20k} and no overall feedback

15.5.1 Circuit diagram and components:

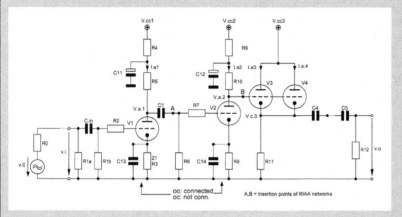

Figure 15.29 Basic pre-amp design without overall feedback and equalization (= Fig. 15.4)

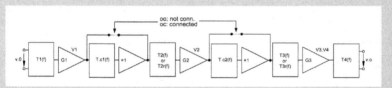

Figure 15.30 Simplified equivalent circuit of Figure 15.29 (= Fig. 15.5)

15.5.1.1   Valve constants and data sheet figures:

V1 ... 4: 1/2 ECC83-12AX7

$I_{a1} = I_{a2} = I_{a3} = I_{a4} = I_a$ $I_a := 1.7 \cdot 10^{-3} A$

$V_{cc1.oc} = V_{cc1.oo} = V_{cc2.oc} = V_{cc2.oo} = 420V$ $V_{cc3} = 420V$

$V_{a.1} = 250V$ $V_{a.2} = 250V$ $V_{a.3} = V_{cc3}$

$g_{m.v1} = g_{m.v2} = g_{m.v3} = g_{m.v4} = g_m$ $g_m := 1.9 \cdot 10^{-3} S$

$\mu_{v1} = \mu_{v2} = \mu_{v3} = \mu_{v4} = \mu$ $\mu := 101$

$r_{a.v1} = r_{a.v2} = r_{a.v3} = r_{a.v4} = r_a$ $r_a := 53 \cdot 10^3 \Omega$

$C_{g.a.83} := 1.6 \cdot 10^{-12} F$ $C_{a.c.83} := 0.33 \cdot 10^{-12} F$ $C_{g.c.83} := 1.65 \cdot 10^{-12} F$

➤ MCD Worksheet XV-2   local feedback versions (flat)   page 2

15.5.1.2   Component figures:

$R0 := 1 \cdot 10^3 \Omega$ $R1a := 52.3 \cdot 10^3 \Omega$ $R1b := 475 \cdot 10^3 \Omega$

                                             $R2 := 100 \Omega$ $R3 := 1 \cdot 10^3 \Omega$

$R6 := 1000 \cdot 10^3 \Omega$ $R7 := 100 \Omega$ $R8 := 1 \cdot 10^3 \Omega$

$R11 := 75 \cdot 10^3 \Omega$ $R12 := 47.5 \cdot 10^3 \Omega$ $R13 := 100 \Omega$

$R5_{oc} := 24.332 \cdot 10^3 \Omega$ $R4_{oc} := 100 \cdot 10^3 \Omega - R5_{oc}$ $R4_{oc} = 75.668 \times 10^3 \Omega$

$R5_{oo} := 15 \cdot 10^3 \Omega$ $R4_{oo} := 100 \cdot 10^3 \Omega - R5_{oo}$ $R4_{oo} = 85 \times 10^3 \Omega$

$R10_{oc} := 100 \cdot 10^3 \Omega$ $R9_{oc} := 100 \cdot 10^3 \Omega - R10_{oc}$ $R9_{oc} = 0 \times 10^0 \Omega$

$R10_{oo} := 16.5 \cdot 10^3 \Omega$ $R9_{oo} := 100 \cdot 10^3 \Omega - R10_{oo}$ $R9_{oo} = 83.5 \times 10^3 \Omega$

$C_{in} := 2.2 \cdot 10^{-6} F$ $C1 := 470 \cdot 10^{-9} F$ $C4 := 22 \cdot 10^{-6} F$

$C5 := 22 \cdot 10^{-6} F$ $C11 = 100 \cdot 10^{-6} F$ $C12 = C11$

$C_{stray.1} := 10 \cdot 10^{-12} F$ $C_{stray.2} := 5 \cdot 10^{-12} F$ $C_{stray.3} := 4 \cdot 10^{-12} F$

15.5.2 Gain calculations

$$G1_{oc} := -\mu \cdot \frac{\left(\dfrac{1}{R5_{oc}} + \dfrac{1}{R6}\right)^{-1}}{r_a + \left(\dfrac{1}{R5_{oc}} + \dfrac{1}{R6}\right)^{-1} + (1 + \mu) \cdot R3}$$ $G1_{oc} = -13.422 \times 10^0$

$$G1_{oo} := -\mu \cdot \frac{\left(\dfrac{1}{R5_{oo}} + \dfrac{1}{R6}\right)^{-1}}{r_a + \left(\dfrac{1}{R5_{oo}} + \dfrac{1}{R6}\right)^{-1}}$$ $G1_{oo} = -22.022 \times 10^0$

$$G2_{oc} := -\mu \cdot \frac{R10_{oc}}{r_a + R10_{oc} + (1 + \mu) \cdot R8}$$ $G2_{oc} = -39.608 \times 10^0$

$$G2_{oo} := -\mu \cdot \frac{R10_{oo}}{r_a + R10_{oo}}$$ $G2_{oo} = -23.978 \times 10^0$

$$r_{a.v3.4} := \frac{r_a}{2}$$

$$G3 := \mu \cdot \frac{R11}{r_{a.v3.4} + (1 + \mu) \cdot R11}$$ $G3 = 986.778 \times 10^{-3}$

$A_{oc} := G1_{oc} \cdot G2_{oc} \cdot G3$

$A_{oc} = 524.57 \times 10^0$

$A_{oo} := G1_{oo} \cdot G2_{oo} \cdot G3$

$A_{oo} = 521.07 \times 10^0$

15.5.3 Network calculations

$f := 10\,Hz, 20\,Hz .. 20000\,Hz$

15.5.3.1   T1:

$C_{i.1.oc.tot} := \left(1 + \left|G1_{oc}\right|\right) \cdot C_{g.a.83} + C_{g.c.83} + C_{stray.1}$

$Z1_1(f) := \dfrac{1}{2j \cdot \pi \cdot f \cdot C_{in}}$

$C_{i.1.oc.tot} = 34.724 \times 10^{-12}\,F$

$Z2_{1.oc}(f) := \dfrac{1}{2j \cdot \pi \cdot f \cdot C_{i.1.oc.tot}}$

$C_{i.1.oo.tot} := \left(1 + \left|G1_{oo}\right|\right) \cdot C_{g.a.83} + C_{g.c.83} + C_{stray.1}$

$C_{i.1.oo.tot} = 48.485 \times 10^{-12}\,F$

$Z2_{1.oo}(f) := \dfrac{1}{2j \cdot \pi \cdot f \cdot C_{i.1.oo.tot}}$

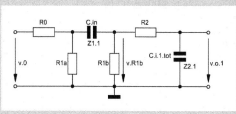

Figure 15.31   T1(f) network (= Fig. 15.10)

$M_{oc}(f) := \dfrac{Z2_{1.oc}(f)}{R2 + Z2_{1.oc}(f)}$

$\left(\dfrac{1}{R1a} + \dfrac{1}{R1b}\right)^{-1} = 47.113 \times 10^3\,\Omega$

$N_{oc}(f) := \dfrac{\left(\dfrac{1}{R1b} + \dfrac{1}{R2 + Z2_{1.oc}(f)}\right)^{-1}}{Z1_1(f) + \left(\dfrac{1}{R1b} + \dfrac{1}{R2 + Z2_{1.oc}(f)}\right)^{-1}}$

$O_{oc}(f) := \dfrac{\left[\dfrac{1}{R1a} + \dfrac{1}{\left[Z1_1(f) + \left(\dfrac{1}{R1b} + \dfrac{1}{R2 + Z2_{1.oc}(f)}\right)^{-1}\right]}\right]^{-1}}{R0 + \left[\dfrac{1}{R1a} + \dfrac{1}{\left[Z1_1(f) + \left(\dfrac{1}{R1b} + \dfrac{1}{R2 + Z2_{1.oc}(f)}\right)^{-1}\right]}\right]^{-1}}$

$T1_{1.oc}(f) := M_{oc}(f) \cdot N_{oc}(f) \cdot O_{oc}(f)$

$$M_{oo}(f) := \frac{Z2_{1.oo}(f)}{R2 + Z2_{1.oo}(f)}$$

$$N_{oo}(f) := \frac{\left(\dfrac{1}{R1b} + \dfrac{1}{R2 + Z2_{1.oo}(f)}\right)^{-1}}{Z1_1(f) + \left(\dfrac{1}{R1b} + \dfrac{1}{R2 + Z2_{1.oo}(f)}\right)^{-1}}$$

$$O_{oo}(f) := \frac{\left[\dfrac{1}{R1a} + \dfrac{1}{Z1_1(f) + \left(\dfrac{1}{R1b} + \dfrac{1}{R2 + Z2_{1.oo}(f)}\right)^{-1}}\right]^{-1}}{R0 + \left[\dfrac{1}{R1a} + \dfrac{1}{Z1_1(f) + \left(\dfrac{1}{R1b} + \dfrac{1}{R2 + Z2_{1.oo}(f)}\right)^{-1}}\right]^{-1}}$$

$$T1_{1.oo}(f) := M_{oo}(f) \cdot N_{oo}(f) \cdot O_{oo}(f)$$

15.5.3.2   Tc1:

$$Tc1_{oc}(f) := 1$$

$$r_{c.1.oo} := \frac{R5_{oo} + r_a}{\mu + 1}$$

$$R_{o.c1.oo} := \left(\frac{1}{r_{c.1.oo}} + \frac{1}{R3}\right)^{-1}$$

$$C13 := \frac{1}{2 \cdot \pi \cdot 0.2Hz R_{o.c1.oo}}$$

$$C13 = 1.989 \times 10^{-3} F$$

$$Tc1_{oo}(f) := \frac{R_{o.c1.oo}}{R_{o.c1.oo} + \left(2j \cdot \pi \cdot f C13\right)^{-1}}$$

$$T1_{oc}(f) := T1_{1.oc}(f) \cdot Tc1_{oc}(f)$$

$$T1_{oo}(f) := T1_{1.oo}(f) \cdot Tc1_{oo}(f)$$

$$T1_{oc.e}(f) := 20 \cdot \log\left(\left|T1_{oc}(f)\right|\right)$$

$$T1_{oo.e}(f) := 20 \cdot \log\left(\left|T1_{oo}(f)\right|\right)$$

$$\phi1_{oc}(f) := \operatorname{atan}\left(\frac{\operatorname{Im}\left(T1_{oc}(f)\right)}{\operatorname{Re}\left(T1_{oc}(f)\right)}\right)$$

$$\phi1_{oo}(f) := \operatorname{atan}\left(\frac{\operatorname{Im}\left(T1_{oo}(f)\right)}{\operatorname{Re}\left(T1_{oo}(f)\right)}\right)$$

➢ MCD Worksheet XV-2 local feedback versions (flat) page 5

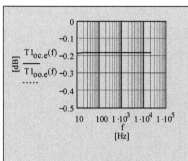

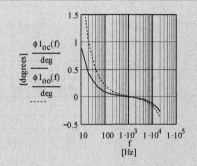

Figure 15.32 Transfers of T1(f)

Figure 15.33 Phases of T1(f)

15.5.3.3 T2:

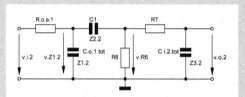

Figure 15.34 T2(f) network
(= Fig. 15.13)

$$R_{o.a.1.oc} := \left[\frac{1}{R5_{oc}} + \frac{1}{r_a + (1 + \mu)R3} \right]^{-1} \qquad R_{o.a.1.oc} = 21.031 \times 10^3 \, \Omega$$

$$R_{o.a.1.oo} := \left[\frac{1}{R5_{oo}} + \frac{1}{r_a + (1 + \mu)R3} \right]^{-1} \qquad R_{o.a.1.oo} = 13.676 \times 10^3 \, \Omega$$

$C_{o.1.oc.tot} := C_{g.a.83} + C_{a.c.83}$ $\qquad C_{i.2.oc.tot} := \left(1 + |G2_{oc}|\right) \cdot C_{g.a.83} + C_{stray.2} + C_{g.c.83}$

$C_{o.1.oc.tot} = 1.93 \times 10^{-12} F$ $\qquad C_{i.2.oc.tot} = 71.623 \times 10^{-12} F$

$C_{o.1.oo.tot} := C_{g.a.83} + C_{a.c.83}$ $\qquad C_{i.2.oo.tot} := \left(1 + |G2_{oo}|\right) \cdot C_{g.a.83} + C_{stray.2} + C_{g.c.83}$

$C_{o.1.oo.tot} = 1.93 \times 10^{-12} F$ $\qquad C_{i.2.oo.tot} = 46.615 \times 10^{-12} F$

$$Z1_{2.oc}(f) := \frac{1}{2j \cdot \pi \cdot f \cdot C_{o.1.oc.tot}} \qquad Z2_{2.oc}(f) := \frac{1}{2j \cdot \pi \cdot f \cdot C1}$$

$$Z1_{2.oo}(f) := \frac{1}{2j \cdot \pi \cdot f \cdot C_{o.1.oo.tot}} \qquad Z2_{2.oo}(f) := \frac{1}{2j \cdot \pi \cdot f \cdot C1}$$

$$Z3_{2.oc}(f) := \frac{1}{2j \cdot \pi \cdot f \cdot C_{i.2.oc.tot}} \qquad Z3_{2.oo}(f) := \frac{1}{2j \cdot \pi \cdot f \cdot C_{i.2.oo.tot}}$$

$$v_{Z1.2}(f) = v_{i.2}(f) \cdot \frac{\left[\dfrac{1}{Z1_2(f)} + \dfrac{1}{Z2_2(f) + \left(\dfrac{1}{Z4_2(f)} + \dfrac{1}{Z3_2(f)} \right)^{-1}} \right]^{-1}}{R_{o.a.1} + \left[\dfrac{1}{Z1_2(f)} + \dfrac{1}{Z2_2(f) + \left(\dfrac{1}{R6} + \dfrac{1}{Z3_2(f)} \right)^{-1}} \right]^{-1}}$$

$$v_{R6}(f) = v_{Z.1.2}(f) \cdot \frac{\left(\dfrac{1}{Z4_2(f)} + \dfrac{1}{R7 + Z3_2(f)} \right)^{-1}}{Z2_2(f) + \left(\dfrac{1}{Z4_2(f)} + \dfrac{1}{R7 + Z3_2(f)} \right)^{-1}}$$

$$v_{o.2}(f) = v_{R6}(f) \cdot \frac{Z3_2(f)}{R7 + Z3_2(f)} \qquad\qquad T2(f) = \frac{v_{o.2}(f)}{v_{1.2}(f)}$$

$$T2_{1.oc}(f) := \frac{\left[\dfrac{1}{Z1_{2.oc}(f)} + \dfrac{1}{Z2_{2.oc}(f) + \left(\dfrac{1}{R6} + \dfrac{1}{Z3_{2.oc}(f)} \right)^{-1}} \right]^{-1}}{R_{o.a.1.oc} + \left[\dfrac{1}{Z1_{2.oc}(f)} + \dfrac{1}{Z2_{2.oc}(f) + \left(\dfrac{1}{R6} + \dfrac{1}{Z3_{2.oc}(f)} \right)^{-1}} \right]^{-1}}$$

$$T2_{1.oo}(f) := \frac{\left[\dfrac{1}{Z1_{2.oo}(f)} + \dfrac{1}{Z2_{2.oo}(f) + \left(\dfrac{1}{R6} + \dfrac{1}{Z3_{2.oo}(f)} \right)^{-1}} \right]^{-1}}{R_{o.a.1.oo} + \left[\dfrac{1}{Z1_{2.oo}(f)} + \dfrac{1}{Z2_{2.oo}(f) + \left(\dfrac{1}{R6} + \dfrac{1}{Z3_{2.oo}(f)} \right)^{-1}} \right]^{-1}}$$

$$T2_{2.oc}(f) := \frac{\left(\dfrac{1}{R6} + \dfrac{1}{R7 + Z3_{2.oc}(f)} \right)^{-1}}{Z2_{2.oc}(f) + \left(\dfrac{1}{R6} + \dfrac{1}{R7 + Z3_{2.oc}(f)} \right)^{-1}} \cdot \frac{Z3_{2.oc}(f)}{R7 + Z3_{2.oc}(f)}$$

➢   MCD Worksheet XV-2   local feedback versions (flat)   page 7

$$T2_{2.oo}(f) := \frac{\left(\dfrac{1}{R6} + \dfrac{1}{R7 + Z3_{2.oo}(f)}\right)^{-1}}{Z2_{2.oo}(f) + \left(\dfrac{1}{R6} + \dfrac{1}{R7 + Z3_{2.oo}(f)}\right)^{-1}} \cdot \frac{Z3_{2.oo}(f)}{R7 + Z3_{2.oo}(f)}$$

15.5.3.4   Tc2 :

$$Tc2_{oc}(f) := 1$$

$$r_{c.2.oo} := \frac{R10_{oo} + r_a}{\mu + 1}$$

$$R_{o.c2.oo} := \left(\frac{1}{r_{c.2.oo}} + \frac{1}{R8}\right)^{-1}$$

$$C14 := \frac{1}{2 \cdot \pi \cdot 0.2 Hz R_{o.c2.oo}}$$

$$C14 = 1.964 \times 10^{-3} F$$

$$Tc2_{oo}(f) := \frac{R_{o.c2.oo}}{R_{o.c2.oo} + \left(2j \cdot \pi \cdot f C14\right)^{-1}}$$

$$T2_{oc}(f) := T2_{1.oc}(f) \cdot T2_{2.oc}(f) \cdot Tc2_{oc}(f)$$

$$T2_{oo}(f) := T2_{1.oo}(f) \cdot T2_{2.oo}(f) \cdot Tc2_{oo}(f)$$

$$T2_{oc.e}(f) := 20 \cdot \log\left(\left|T2_{oc}(f)\right|\right)$$

$$T2_{oo.e}(f) := 20 \cdot \log\left(\left|T2_{oo}(f)\right|\right)$$

$$\phi2_{oc}(f) := atan\left(\frac{Im\left(T2_{oc}(f)\right)}{Re\left(T2_{oc}(f)\right)}\right)$$

$$\phi2_{oo}(f) := atan\left(\frac{Im\left(T2_{oo}(f)\right)}{Re\left(T2_{oo}(f)\right)}\right)$$

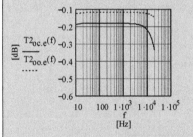

Figure 15.35 Transfers of T2(f)

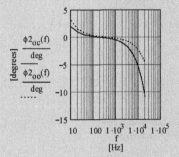

Figure 15.36 Phases of T2(f)

➢   MCD Worksheet XV-2   local feedback versions (flat)   page 8

15.5.3.5   T3:

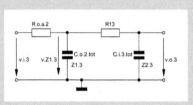

Figure 15.37 T3(f) network (= Fig. 15.16)

$$T3(f) = \frac{v_{o.3}(f)}{v_{i.3}(f)} \qquad\qquad v_{o.3}(f) = v_{Z.1.3}(f) \cdot \frac{Z2_3(f)}{Z2_3(f) + R13}$$

$$v_{Z1.3}(f) = v_{i.3}(f) \cdot \frac{\left(\dfrac{1}{Z1_3(f)} + \dfrac{1}{R13 + Z2_3(f)}\right)^{-1}}{R_{o.a.2} + \left(\dfrac{1}{Z1_3(f)} + \dfrac{1}{R13 + Z2_3(f)}\right)^{-1}}$$

$$R_{o.a.2.oc} := \left[\frac{1}{R10_{oc}} + \frac{1}{r_a + (1 + \mu)R8}\right]^{-1} \qquad R_{o.a.2.oc} = 60.784 \times 10^3\,\Omega$$

$$R_{o.a.2.oo} := \left[\frac{1}{R10_{oo}} + \frac{1}{r_a + (1 + \mu)R8}\right]^{-1} \qquad R_{o.a.2.oo} = 14.913 \times 10^3\,\Omega$$

$$C_{i.3.tot} := C_{g.a.83} + C_{stray.3} + (1 - |G3|) \cdot C_{g.c.83} \qquad C_{i.3.tot} = 5.622 \times 10^{-12}\,F$$

$$C_{o.2.tot} := C_{a.c.83} + C_{g.a.83} \qquad\qquad C_{o.2.tot} = 1.93 \times 10^{-12}\,F$$

$$C_{i.3.tot} := C_{g.a.83} + C_{stray.3} + (1 - |G3|) \cdot C_{g.c.83} \qquad C_{i.3.tot} = 5.622 \times 10^{-12}\,F$$

$$C_{o.2.oo.tot} := C_{a.c.83} + C_{g.a.83} \qquad\qquad C_{o.2.oo.tot} = 1.93 \times 10^{-12}\,F$$

$$Z1_3(f) := \frac{1}{2j \cdot \pi \cdot f \cdot C_{o.2.tot}} \qquad\qquad Z2_3(f) := \frac{1}{2j \cdot \pi \cdot f \cdot C_{i.3.tot}}$$

$$Z1_{3.oo}(f) := \frac{1}{2j \cdot \pi \cdot f \cdot C_{o.2.oo.tot}} \qquad\qquad Z2_3(f) := \frac{1}{2j \cdot \pi \cdot f \cdot C_{i.3.tot}}$$

$$T3_{oc}(f) := \frac{\left(\dfrac{1}{Z1_3(f)} + \dfrac{1}{R13 + Z2_3(f)}\right)^{-1}}{R_{o.a.2.oc} + \left(\dfrac{1}{Z1_3(f)} + \dfrac{1}{R13 + Z2_3(f)}\right)^{-1}} \cdot \frac{Z2_3(f)}{Z2_3(f) + R13}$$

➢ MCD Worksheet XV-2   local feedback versions (flat)   page 9

$$T3_{oo}(f) := \cfrac{\left(\cfrac{1}{Zl_{3.oo}(f)} + \cfrac{1}{R13 + Z2_3(f)}\right)^{-1}}{R_{o.a.2.oo} + \left(\cfrac{1}{Zl_{3.oo}(f)} + \cfrac{1}{R13 + Z2_3(f)}\right)^{-1}} \cdot \cfrac{Z2_3(f)}{Z2_3(f) + R13}$$

$$T3_{oc.e}(f) := 20 \cdot \log\left(\left|T3_{oc}(f)\right|\right) \qquad \phi3_{oc}(f) := atan\left(\cfrac{Im\left(T3_{oc}(f)\right)}{Re\left(T3_{oc}(f)\right)}\right)$$

$$T3_{oo.e}(f) := 20 \cdot \log\left(\left|T3_{oo}(f)\right|\right) \qquad \phi3_{oo}(f) := atan\left(\cfrac{Im\left(T3_{oo}(f)\right)}{Re\left(T3_{oo}(f)\right)}\right)$$

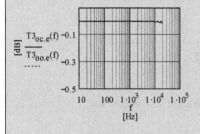

Figure 15.38 Transfers of T3(f)

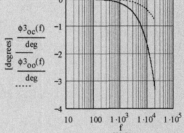

Figure 15.39 Phases of T3(f)

15.5.3.6   T4:

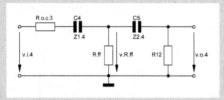

Figure 15.40 T4(f) network (= Fig. 15.19)

$$r_{c.3} := \cfrac{r_{a.v3.4}}{\mu + 1} \qquad\qquad r_{c.3} = 259.804 \times 10^0 \, \Omega$$

$$R_{o.c.3} := \left(\cfrac{1}{r_{c.3}} + \cfrac{1}{R11}\right)^{-1} \qquad\qquad R_{o.c.3} = 258.907 \times 10^0 \, \Omega$$

➢   MCD Worksheet XV-2   local feedback versions (flat)   page 10

$$Z1_4(f) := \frac{1}{2j \cdot \pi \cdot f \cdot C4}$$

$$Z2_4(f) := \frac{1}{2j \cdot \pi \cdot f \cdot C5}$$

$$T4(f) = \frac{v_{o.4}(f)}{v_{i.4}(f)}$$

$$T4(f) := \frac{R12}{R_{o.c.3} + R12 + Z1_4(f) + Z2_4(f)}$$

$$T4_e(f) := 20 \cdot \log(|T4(f)|)$$

$$\phi4(f) := atan\left(\frac{Im(T4(f))}{Re(T4(f))}\right)$$

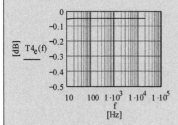

Figure 15.41 Transfer of T4(f)

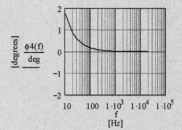

Figure 15.42 Phase of T4(f)

15.5.4 Total gains A $_{xy.tot}$(f) and phases ϕ $_{xy.tot}$(f) for the "flat" pre-amps:

$$A_{oc.tot}(f) := T1_{oc}(f) \cdot T2_{oc}(f) \cdot T3_{oc}(f) \cdot T4(f) \cdot A_{oc}$$

$$A_{oc.tot.e}(f) := 20 \cdot \log(|A_{oc.tot}(f)|)$$

$$A_{oo.tot}(f) := T1_{oo}(f) \cdot T2_{oo}(f) \cdot T3_{oo}(f) \cdot T4(f) \cdot A_{oo}$$

$$A_{oo.tot.e}(f) := 20 \cdot \log(|A_{oo.tot}(f)|)$$

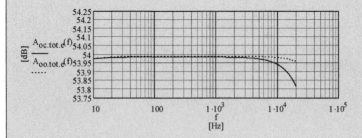

Figure 15.43 Frequency responses of both pre-amps

➢   MCD Worksheet XV-2   local feedback versions (flat)   page 11

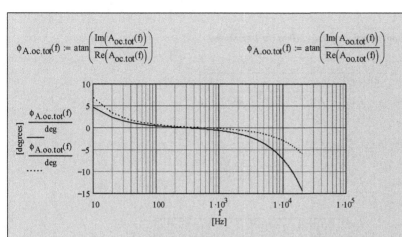

$$\phi_{A.oc.tot}(f) := atan\left(\frac{Im\left(A_{oc.tot}(f)\right)}{Re\left(A_{oc.tot}(f)\right)}\right) \qquad \phi_{A.oo.tot}(f) := atan\left(\frac{Im\left(A_{oo.tot}(f)\right)}{Re\left(A_{oo.tot}(f)\right)}\right)$$

Figure 15.44 Phases responses of both pre-amps

$$A_{oc.tot.e}\left(10^3 Hz\right) = 53.984 \times 10^0 \qquad\qquad \frac{\phi_{A.oc.tot}(20Hz)}{deg} = 2.239 \times 10^0$$

$$A_{oo.tot.e}\left(10^3 Hz\right) = 53.989 \times 10^0 \qquad\qquad \frac{\phi_{A.oo.tot}(20Hz)}{deg} = 3.4 \times 10^0$$

$$A_{oc.tot.e}\left(20 \cdot 10^3 Hz\right) = 53.815 \times 10^0 \qquad\qquad \frac{\phi_{A.oc.tot}\left(20 \cdot 10^3 Hz\right)}{deg} = -14.402 \times 10^0$$

$$A_{oo.tot.e}\left(20 \cdot 10^3 Hz\right) = 53.959 \times 10^0 \qquad\qquad \frac{\phi_{A.oo.tot}\left(20 \cdot 10^3 Hz\right)}{deg} = -5.926 \times 10^0$$

15.5.5 Input impedances $Z_{in.A.xy}(f)$ of the 2 versions:

$$Z_{in.A.oc}(f) := \left[\frac{1}{R1a} + \cfrac{1}{Z1_1(f) + \left(\frac{1}{R1b} + \cfrac{1}{R2 + Z2_{1.oc}(f)}\right)^{-1}}\right]^{-1}$$

$$Z_{in.A.oo}(f) := \left[\frac{1}{R1a} + \cfrac{1}{Z1_1(f) + \left(\frac{1}{R1b} + \cfrac{1}{R2 + Z2_{1.oo}(f)}\right)^{-1}}\right]^{-1}$$

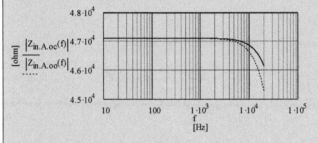

Figure 15.45 Input impedances of the two versions

➤ MCD Worksheet XV-3   local feedback versions (RIAA)   page 1

15.6   Example calculations for a 3-stage MM phono pre-amp ($B_{oc}+B_{oo}$ versions)

15.6.1 Circuit diagram and components:

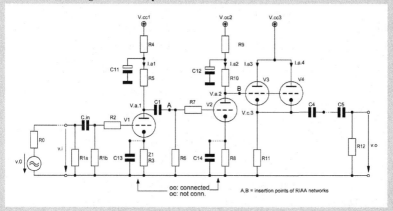

Figure 15.46 Basic pre-amp design with passive RIAA equalization and without overall feedback (= Fig. 15.4)

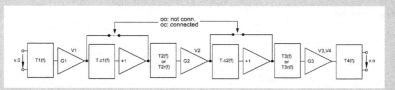

Figure 15.47 Simplified equivalent circuit of Figure 15.46 (= Fig. 15.5)

15.6.1.1   Valve constants and data sheet figures:

V1 ... 4: 1/2 ECC83-12AX7

$I_{a1} = I_{a2} = I_{a3} = I_{a4} = I_a$ $I_a := 1.7 \cdot 10^{-3} A$

$V_{cc1.oc} = V_{cc1.oo} = V_{cc2.oc} = V_{cc2.oo} = 420V$ $V_{cc3} = 420V$

$V_{a.1} = 250V$ $V_{a.2} = 250V$ $V_{a.3} = V_{cc3}$

$g_{m.v1} = g_{m.v2} = g_{m.v3} = g_{m.v4} = g_m$ $g_m := 1.9 \cdot 10^{-3} S$

$\mu_{v1} = \mu_{v2} = \mu_{v3} = \mu_{v4} = \mu$ $\mu := 101$

$r_{a.v1} = r_{a.v2} = r_{a.v3} = r_{a.v4} = r_a$ $r_a := 53 \cdot 10^3 \Omega$

$C_{g.a.83} := 1.6 \cdot 10^{-12} F$ $C_{a.c.83} := 0.33 \cdot 10^{-12} F$ $C_{g.c.83} := 1.65 \cdot 10^{-12} F$

➢   **MCD Worksheet XV-3**   local feedback versions (RIAA)   page 2

15.6.1.2   Component figures:

$R0 := 1 \cdot 10^3 \Omega$ $R1a := 52.3 \cdot 10^3 \Omega$ $R1b := 475 \cdot 10^3 \Omega$

                                                $R2 := 100 \Omega$ $R3 := 1 \cdot 10^3 \Omega$

$R6 := 1000 \cdot 10^3 \Omega$ $R7 := 100 \Omega$ $R8 := 1 \cdot 10^3 \Omega$

$R11 := 75 \cdot 10^3 \Omega$ $R12 := 47.5 \cdot 10^3 \Omega$ $R13 := 0.1 \Omega$

$R5_{oc} := 47 \cdot 10^3 \Omega$ $R4_{oc} := 100 \cdot 10^3 \Omega - R5_{oc}$ $R4_{oc} = 53 \times 10^3 \Omega$

$R5_{oo} := 15 \cdot 10^3 \Omega$ $R4_{oo} := 100 \cdot 10^3 \Omega - R5_{oo}$ $R4_{oo} = 85 \times 10^3 \Omega$

$R10_{oc} := 47.5 \cdot 10^3 \Omega$ $R9_{oc} := 100 \cdot 10^3 \Omega - R10_{oc}$ $R9_{oc} = 52.5 \times 10^3 \Omega$

$R10_{oo} := 16.5 \cdot 10^3 \Omega$ $R9_{oo} := 100 \cdot 10^3 \Omega - R10_{oo}$ $R9_{oo} = 83.5 \times 10^3 \Omega$

$C_{in} := 2.2 \cdot 10^{-6} F$ $C1 := 470 \cdot 10^{-9} F$ $C4 := 22 \cdot 10^{-6} F$

$C5 := 22 \cdot 10^{-6} F$ $C11 := 10 \cdot 10^{-6} F$ $C12 = C11$

$C_{stray.1} := 10 \cdot 10^{-12} F$ $C_{stray.2} := 5 \cdot 10^{-12} F$ $C_{stray.3} := 4 \cdot 10^{-12} F$

15.6.2 Gain calculations

$$G1_{oc} := -\mu \cdot \frac{\left(\dfrac{1}{R5_{oc}} + \dfrac{1}{R6}\right)^{-1}}{r_a + \left(\dfrac{1}{R5_{oc}} + \dfrac{1}{R6}\right)^{-1} + (1 + \mu) \cdot R3}$$ $G1_{oc} = -22.682 \times 10^0$

$$G1_{oo} := -\mu \cdot \frac{\left(\dfrac{1}{R5_{oo}} + \dfrac{1}{R6}\right)^{-1}}{r_a + \left(\dfrac{1}{R5_{oo}} + \dfrac{1}{R6}\right)^{-1}}$$ $G1_{oo} = -22.022 \times 10^0$

$$G2_{oc} := -\mu \cdot \frac{R10_{oc}}{r_a + R10_{oc} + (1 + \mu) \cdot R8}$$ $G2_{oc} = -23.691 \times 10^0$

$$G2_{oo} := -\mu \cdot \frac{R10_{oo}}{r_a + R10_{oo}}$$ $G2_{oo} = -23.978 \times 10^0$

$$r_{a.v3.4} := \frac{r_a}{2}$$

$$G3 := \mu \cdot \frac{R11}{r_{a.v3.4} + (1 + \mu) \cdot R11}$$ $G3 = 986.778 \times 10^{-3}$

$B_{oc} := G1_{oc} \cdot G2_{oc} \cdot G3$ $B_{oc} = 530.262 \times 10^0$

$B_{oo} := G1_{oo} \cdot G2_{oo} \cdot G3$ $B_{oo} = 521.07 \times 10^0$

15.6.3 Network calculations

$f := 10\,Hz, 20\,Hz .. 20000\,Hz$

$f_{opt} := 0.2\,Hz$

15.6.3.1   T1:

$C_{i.1.oc.tot} := \left(1 + \left|G1_{oc}\right|\right) \cdot C_{g.a.83} + C_{g.c.83} + C_{stray.1}$ $Z1_1(f) := \dfrac{1}{2j \cdot \pi \cdot f \cdot C_{in}}$

$C_{i.1.oc.tot} = 49.541 \times 10^{-12}\,F$ $Z2_{1.oc}(f) := \dfrac{1}{2j \cdot \pi \cdot f \cdot C_{i.1.oc.tot}}$

$C_{i.1.oo.tot} := \left(1 + \left|G1_{oo}\right|\right) \cdot C_{g.a.83} + C_{g.c.83} + C_{stray.1}$

$C_{i.1.oo.tot} = 48.485 \times 10^{-12}\,F$ $Z2_{1.oo}(f) := \dfrac{1}{2j \cdot \pi \cdot f \cdot C_{i.1.oo.tot}}$

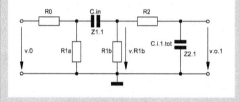

Figure 15.48 T1(f) network
(= Fig. 15.10)

$M_{oc}(f) := \dfrac{Z2_{1.oc}(f)}{R2 + Z2_{1.oc}(f)}$ $\left(\dfrac{1}{R1a} + \dfrac{1}{R1b}\right)^{-1} = 47.113 \times 10^3\,\Omega$

$N_{oc}(f) := \dfrac{\left(\dfrac{1}{R1b} + \dfrac{1}{R2 + Z2_{1.oc}(f)}\right)^{-1}}{Z1_1(f) + \left(\dfrac{1}{R1b} + \dfrac{1}{R2 + Z2_{1.oc}(f)}\right)^{-1}}$

$O_{oc}(f) := \dfrac{\left[\dfrac{1}{R1a} + \dfrac{1}{\left[Z1_1(f) + \left(\dfrac{1}{R1b} + \dfrac{1}{R2 + Z2_{1.oc}(f)}\right)^{-1}\right]}\right]^{-1}}{R0 + \left[\dfrac{1}{R1a} + \dfrac{1}{\left[Z1_1(f) + \left(\dfrac{1}{R1b} + \dfrac{1}{R2 + Z2_{1.oc}(f)}\right)^{-1}\right]}\right]^{-1}}$

$T1_{1.oc}(f) := M_{oc}(f) \cdot N_{oc}(f) \cdot O_{oc}(f)$

> MCD Worksheet XV-3   local feedback versions (RIAA)   page 4

$$M_{oo}(f) := \frac{Z2_{1.oo}(f)}{R2 + Z2_{1.oo}(f)}$$

$$N_{oo}(f) := \frac{\left(\dfrac{1}{R1b} + \dfrac{1}{R2 + Z2_{1.oo}(f)}\right)^{-1}}{Z1_1(f) + \left(\dfrac{1}{R1b} + \dfrac{1}{R2 + Z2_{1.oo}(f)}\right)^{-1}}$$

$$O_{oo}(f) := \frac{\left[\dfrac{1}{R1a} + \dfrac{1}{\left[Z1_1(f) + \left(\dfrac{1}{R1b} + \dfrac{1}{R2 + Z2_{1.oo}(f)}\right)^{-1}\right]}\right]^{-1}}{R0 + \left[\dfrac{1}{R1a} + \dfrac{1}{\left[Z1_1(f) + \left(\dfrac{1}{R1b} + \dfrac{1}{R2 + Z2_{1.oo}(f)}\right)^{-1}\right]}\right]^{-1}}$$

$$T1_{1.oo}(f) := M_{oo}(f) \cdot N_{oo}(f) \cdot O_{oo}(f)$$

15.6.3.2   Tc1:

$$Tc1_{oc}(f) := 1$$

$$r_{c.1.oo} := \frac{R5_{oo} + r_a}{\mu + 1} \qquad\qquad\qquad R_{o.c1.oo} := \left(\frac{1}{r_{c.1.oo}} + \frac{1}{R3}\right)^{-1}$$

$$C13 := \frac{1}{2 \cdot \pi \cdot f_{opt} \cdot R_{o.c1.oo}} \qquad\qquad\qquad C13 = 1.989 \times 10^{-3}\,F$$

$$Tc1_{oo}(f) := \frac{R_{o.c1.oo}}{R_{o.c1.oo} + (2j \cdot \pi \cdot f \cdot C13)^{-1}}$$

$$T1_{oc}(f) := T1_{1.oc}(f) \cdot Tc1_{oc}(f) \qquad\qquad\qquad T1_{oo}(f) := T1_{1.oo}(f) \cdot Tc1_{oo}(f)$$

$$T1_{oc.e}(f) := 20 \cdot \log\left(\left|T1_{oc}(f)\right|\right) \qquad\qquad T1_{oo.e}(f) := 20 \cdot \log\left(\left|T1_{oo}(f)\right|\right)$$

$$\phi1_{oc}(f) := \operatorname{atan}\left(\frac{\operatorname{Im}\left(T1_{oc}(f)\right)}{\operatorname{Re}\left(T1_{oc}(f)\right)}\right) \qquad\qquad \phi1_{oo}(f) := \operatorname{atan}\left(\frac{\operatorname{Im}\left(T1_{oo}(f)\right)}{\operatorname{Re}\left(T1_{oo}(f)\right)}\right)$$

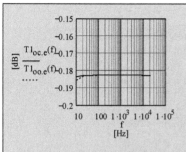

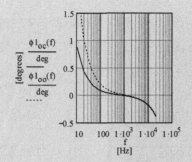

Figure 15.49 Transfers of T1(f)

Figure 15.50 Phases of T1(f)

15.6.3.3   T2.r:

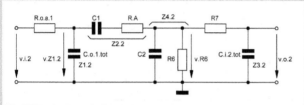

Figure 15.51 T2(f) network for the passive RIAA transfer solution
(including 75µs time constant)

$$R_{o.a.1.\infty} := \left[\frac{1}{R5_{\infty}} + \frac{1}{r_a + (1 + \mu)R3} \right]^{-1} \qquad R_{o.a.1.\infty} = 36.064 \times 10^3\,\Omega$$

$$R_{o.a.1.oo} := \left[\frac{1}{R5_{oo}} + \frac{1}{r_a + (1 + \mu)R3} \right]^{-1} \qquad R_{o.a.1.oo} = 13.676 \times 10^3\,\Omega$$

$$C_{o.1.\infty.tot} := C_{g.a.83} + C_{a.c.83} \qquad\qquad C_{i.2.\infty.tot} := \left(1 + \left|G2_{\infty}\right|\right) \cdot C_{g.a.83} + C_{stray.2} + C_{g.c.83}$$

$$C_{o.1.\infty.tot} = 1.93 \times 10^{-12}\,F \qquad\qquad C_{i.2.\infty.tot} = 46.156 \times 10^{-12}\,F$$

$$C_{o.1.oo.tot} := C_{g.a.83} + C_{a.c.83} \qquad\qquad C_{i.2.oo.tot} := \left(1 + \left|G2_{oo}\right|\right) \cdot C_{g.a.83} + C_{stray.2} + C_{g.c.83}$$

$$C_{o.1.oo.tot} = 1.93 \times 10^{-12}\,F \qquad\qquad C_{i.2.oo.tot} = 46.615 \times 10^{-12}\,F$$

By varying R_A and by keeping C2 fixed at a certain value we'll get app. the 75µs time
constant. The final time constant can be determined with the help of Figure 15.60 :

$$R_{A.\infty} := 6.34 \cdot 10^3\,\Omega \qquad C2_{\infty} := 1.8 \cdot 10^{-9}\,F$$

> MCD Worksheet XV-3   local feedback versions (RIAA)   page 6

$$\left(C2_{oc} + C_{i.2.oc.tot}\right)\left(\frac{1}{R_{o.a.1.oc} + R_{A.oc}} + \frac{1}{R6}\right)^{-1} = 75.1 \times 10^{-6}\,s$$

$$R_{A.oo} := 6.2 \cdot 10^3\,\Omega \qquad C2_{oo} := 3.7 \cdot 10^{-9}\,F$$

$$\left(C2_{oo} + C_{i.2.oo.tot}\right)\cdot\left(\frac{1}{R_{o.a.1.oo} + R_{A.oo}} + \frac{1}{R6}\right)^{-1} = 73.018 \times 10^{-6}\,s$$

$$Z1_{2.oc}(f) := \frac{1}{2j \cdot \pi \cdot f C_{o.1.oc.tot}} \qquad\qquad Z2_{2.oc}(f) := \frac{1}{2j \cdot \pi \cdot f C1} + R_{A.oc}$$

$$Z1_{2.oo}(f) := \frac{1}{2j \cdot \pi \cdot f C_{o.1.oo.tot}} \qquad\qquad Z2_{2.oo}(f) := \frac{1}{2j \cdot \pi \cdot f C1} + R_{A.oo}$$

$$Z3_{2.oc}(f) := \frac{1}{2j \cdot \pi \cdot f C_{i.2.oc.tot}} \qquad\qquad Z4_{2.oc}(f) := \left(2j \cdot \pi \cdot f C2_{oc} + \frac{1}{R6}\right)^{-1}$$

$$Z3_{2.oo}(f) := \frac{1}{2j \cdot \pi \cdot f C_{i.2.oo.tot}} \qquad\qquad Z4_{2.oo}(f) := \left(2j \cdot \pi \cdot f C2_{oo} + \frac{1}{R6}\right)^{-1}$$

$$T2_{1.oc.r}(f) := \frac{\left[\dfrac{1}{Z1_{2.oc}(f)} + \dfrac{1}{Z2_{2.oc}(f) + \left(\dfrac{1}{Z4_{2.oc}(f)} + \dfrac{1}{Z3_{2.oc}(f)}\right)^{-1}}\right]^{-1}}{R_{o.a.1.oc} + \left[\dfrac{1}{Z1_{2.oc}(f)} + \dfrac{1}{Z2_{2.oc}(f) + \left(\dfrac{1}{Z4_{2.oc}(f)} + \dfrac{1}{Z3_{2.oc}(f)}\right)^{-1}}\right]^{-1}}$$

$$T2_{1.oo.r}(f) := \frac{\left[\dfrac{1}{Z1_{2.oo}(f)} + \dfrac{1}{Z2_{2.oo}(f) + \left(\dfrac{1}{Z4_{2.oo}(f)} + \dfrac{1}{Z3_{2.oo}(f)}\right)^{-1}}\right]^{-1}}{R_{o.a.1.oo} + \left[\dfrac{1}{Z1_{2.oo}(f)} + \dfrac{1}{Z2_{2.oo}(f) + \left(\dfrac{1}{Z4_{2.oo}(f)} + \dfrac{1}{Z3_{2.oo}(f)}\right)^{-1}}\right]^{-1}}$$

$$T2_{2.oc.r}(f) := \frac{\left(\dfrac{1}{Z4_{2.oc}(f)} + \dfrac{1}{R7 + Z3_{2.oc}(f)}\right)^{-1}}{Z2_{2.oc}(f) + \left(\dfrac{1}{Z4_{2.oc}(f)} + \dfrac{1}{R7 + Z3_{2.oc}(f)}\right)^{-1}} \cdot \frac{Z3_{2.oc}(f)}{R7 + Z3_{2.oc}(f)}$$

$$T2_{2.oo.r}(f) := \frac{\left(\dfrac{1}{Z4_{2.oo}(f)} + \dfrac{1}{R7 + Z3_{2.oo}(f)}\right)^{-1}}{Z2_{2.oo}(f) + \left(\dfrac{1}{Z4_{2.oo}(f)} + \dfrac{1}{R7 + Z3_{2.oo}(f)}\right)^{-1}} \cdot \frac{Z3_{2.oo}(f)}{R7 + Z3_{2.oo}(f)}$$

15.6.3.4   Tc2 :

$$Tc2_{oc}(f) := 1$$

$$r_{c.2.oo} := \frac{R10_{oo} + r_a}{\mu + 1}$$

$$R_{o.c2.oo} := \left(\frac{1}{r_{c.2.oo}} + \frac{1}{R8} \right)^{-1}$$

$$C14 := \frac{1}{2 \cdot \pi \cdot f_{opt} \cdot R_{o.c2.oo}}$$

$$C14 = 1.964 \times 10^{-3} F$$

$$Tc2_{oo}(f) := \frac{R_{o.c2.oo}}{R_{o.c2.oo} + (2j \cdot \pi \cdot f \cdot C14)^{-1}}$$

$$T2_{oc.r}(f) := T2_{1.oc.r}(f) \cdot T2_{2.oc.r}(f) \cdot Tc2_{oc}(f)$$

$$T2_{oc.r.e}(f) := 20 \cdot \log \left(\left| T2_{oc.r}(f) \right| \right)$$

$$T2_{oo.r}(f) := T2_{1.oo.r}(f) \cdot T2_{2.oo.r}(f) \cdot Tc2_{oo}(f)$$

$$T2_{oo.r.e}(f) := 20 \cdot \log \left(\left| T2_{oo.r}(f) \right| \right)$$

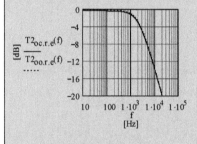

Figure 15.52 T2(f) transfer plots for the passive RIAA transfer solution

15.6.3.5   T3.r:

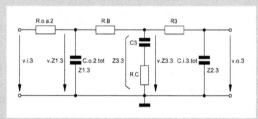

Figure 15.53 T3(f) network for the passive RIAA transfer solution
(including time constants 318 µs and 3180 µs)

$$R_{o.a.2.oc} := \left[\frac{1}{R10_{oc}} + \frac{1}{r_a + (1 + \mu)R8} \right]^{-1}$$

$$R_{o.a.2.oc} = 36.358 \times 10^3 \, \Omega$$

➢   MCD Worksheet XV-3   local feedback versions (RIAA)   page 8

$$R_{o.a.2.oo} := \left[\frac{1}{R10_{oo}} + \frac{1}{r_a + (1 + \mu)R8} \right]^{-1} \qquad R_{o.a.2.oo} = 14.913 \times 10^3 \, \Omega$$

$$C_{i.3.tot} := C_{g.a.83} + C_{stray.3} + (1 - |G3|) \cdot C_{g.c.83} \qquad C_{i.3.tot} = 5.622 \times 10^{-12} F$$

$$C_{o.2.oc.tot} := C_{a.c.83} + C_{g.a.83} \qquad C_{o.2.oc.tot} = 1.93 \times 10^{-12} F$$

$$C_{o.2.oo.tot} := C_{o.2.oc.tot}$$

By varying R_B and keeping C3 and R_C fixed at certain values we'll get app. the time constants $318\,\mu s$ and $3180\,\mu s$. The final time constants can be determined with the help of Figure 15.60 :

$$R_{B.oc} := 249 \cdot 10^3 \Omega + 0.825 \cdot 10^3 \Omega \qquad R_C := 31.6 \cdot 10^3 \Omega + 200\Omega \qquad C3 := 10 \cdot 10^{-9} F$$

$$R_{B.oo} := 270 \cdot 10^3 \Omega + 1.3 \cdot 10^3 \Omega \qquad\qquad C3 \cdot R_C = 318 \times 10^{-6} s$$

$$C3 \cdot \left(R_{B.oc} + R_C + R_{o.a.2.oc} \right) = 3.18 \times 10^{-3} s$$

$$C3 \cdot \left(R_{B.oo} + R_C + R_{o.a.2.oo} \right) = 3.18 \times 10^{-3} s$$

$$Z1_{3.oc}(f) := \frac{1}{2j \cdot \pi \cdot f C_{o.2.oc.tot}} \qquad\qquad Z2_3(f) := \frac{1}{2j \cdot \pi \cdot f C_{i.3.tot}}$$

$$Z1_{3.oo}(f) := \frac{1}{2j \cdot \pi \cdot f C_{o.2.oo.tot}} \qquad\qquad Z2_3(f) := \frac{1}{2j \cdot \pi \cdot f C_{i.3.tot}}$$

$$Z3_3(f) := R_C + \frac{1}{2j \cdot \pi \cdot f C3}$$

$$T3(f) = \frac{v_{o.3}(f)}{v_{i.3}(f)} \qquad\qquad v_{o.3}(f) = v_{Z3.3}(f) \cdot \frac{Z2_3(f)}{Z2_3(f) + R3}$$

$$v_{Z3.3}(f) = v_{Z1.3}(f) \cdot \frac{\left(\frac{1}{Z3_3(f)} + \frac{1}{R3 + Z2_3(f)} \right)^{-1}}{R_B + \left(\frac{1}{Z3_3(f)} + \frac{1}{R3 + Z2_3(f)} \right)^{-1}}$$

$$v_{Z.1.3}(f) = v_{i.3}(f) \cdot \frac{\left[2j \cdot \pi \cdot f C_{o.2.tot} + \left[R_B + \left(\frac{1}{Z3_3(f)} + \frac{1}{R3 + Z2_3(f)} \right)^{-1} \right]^{-1} \right]^{-1}}{R_{o.a.2} + \left[2j \cdot \pi \cdot f C_{o.2.tot} + \left[R_B + \left(\frac{1}{Z3_3(f)} + \frac{1}{R3 + Z2_3(f)} \right)^{-1} \right]^{-1} \right]^{-1}}$$

➤   MCD Worksheet XV-3   local feedback versions (RIAA)   page 9

$$P(f) := \frac{Z2_3(f)}{Z2_3(f) + R3}$$

$$Q_{oc}(f) := \frac{\left(\dfrac{1}{Z3_3(f)} + \dfrac{1}{R3 + Z2_3(f)}\right)^{-1}}{R_{B.oc} + \left(\dfrac{1}{Z3_3(f)} + \dfrac{1}{R3 + Z2_3(f)}\right)^{-1}} \qquad Q_{oo}(f) := \frac{\left(\dfrac{1}{Z3_3(f)} + \dfrac{1}{R3 + Z2_3(f)}\right)^{-1}}{R_{B.oo} + \left(\dfrac{1}{Z3_3(f)} + \dfrac{1}{R3 + Z2_3(f)}\right)^{-1}}$$

$$R_{oc}(f) := \frac{\left[2j\cdot\pi\cdot f\cdot C_{o.2.oc.tot} + \left[R_{B.oc} + \left(\dfrac{1}{Z3_3(f)} + \dfrac{1}{R3 + Z2_3(f)}\right)^{-1}\right]^{-1}\right]^{-1}}{R_{o.a.2.oc} + \left[2j\cdot\pi\cdot f\cdot C_{o.2.oc.tot} + \left[R_{B.oc} + \left(\dfrac{1}{Z3_3(f)} + \dfrac{1}{R3 + Z2_3(f)}\right)^{-1}\right]^{-1}\right]^{-1}}$$

$$R_{oo}(f) := \frac{\left[2j\cdot\pi\cdot f\cdot C_{o.2.oo.tot} + \left[R_{B.oo} + \left(\dfrac{1}{Z3_3(f)} + \dfrac{1}{R3 + Z2_3(f)}\right)^{-1}\right]^{-1}\right]^{-1}}{R_{o.a.2.oo} + \left[2j\cdot\pi\cdot f\cdot C_{o.2.oo.tot} + \left[R_{B.oo} + \left(\dfrac{1}{Z3_3(f)} + \dfrac{1}{R3 + Z2_3(f)}\right)^{-1}\right]^{-1}\right]^{-1}}$$

$$T3_{oc.r}(f) := P(f)\cdot Q_{oc}(f)\cdot R_{oc}(f) \qquad\qquad T3_{oo.r}(f) := P(f)\cdot Q_{oo}(f)\cdot R_{oo}(f)$$

$$T3_{oc.r.e}(f) := 20\cdot\log\left(\left|T3_{oc.r}(f)\right|\right) \qquad\qquad T3_{oo.r.e}(f) := 20\cdot\log\left(\left|T3_{oo.r}(f)\right|\right)$$

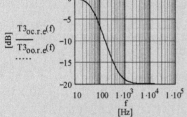

Figure 15.54 T3(f) transfer plots for
the passive RIAA solution

15.6.3.6   T4:

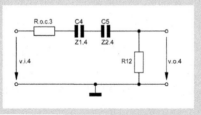

Figure 15.55 T4(f) network for the passive
RIAA transfer solution

$$r_{c.3} := \frac{r_{a.v3.4}}{\mu + 1}$$

$$r_{c.3} = 259.804 \times 10^0 \, \Omega$$

$$R_{o.c.3} := \left(\frac{1}{r_{c.3}} + \frac{1}{R11} \right)^{-1}$$

$$R_{o.c.3} = 258.907 \times 10^0 \, \Omega$$

$$Z1_4(f) := \frac{1}{2j \cdot \pi \cdot f \cdot C4}$$

$$Z2_4(f) := \frac{1}{2j \cdot \pi \cdot f \cdot C5}$$

$$T4(f) = \frac{v_{o.4}(f)}{v_{i.4}(f)}$$

$$T4(f) := \frac{R12}{R_{o.c.3} + R12 + Z1_4(f) + Z2_4(f)}$$

$$T4_e(f) := 20 \cdot \log(|T4(f)|)$$

$$\phi4(f) := \operatorname{atan}\left(\frac{\operatorname{Im}(T4(f))}{\operatorname{Re}(T4(f))} \right)$$

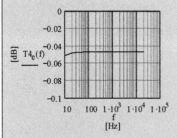

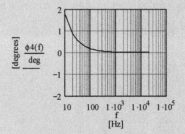

Figure 15.56 Transfer of T4(f)   Figure 15.57 Phase of T4(f)

15.6.4 Total gains $B_{xy.rtot}(f)$ of both pre-amp versions

$$B_{oc.r.tot}(f) := T1_{oc}(f) \cdot T2_{oc.r}(f) \cdot T3_{oc.r}(f) \cdot T4(f) \cdot B_{oc} \qquad B_{oc.r.tot.e}(f) := 20 \cdot \log\left(\left|B_{oc.r.tot}(f)\right|\right)$$

$$B_{oo.r.tot}(f) := T1_{oo}(f) \cdot T2_{oo.r}(f) \cdot T3_{oo.r}(f) \cdot T4(f) \cdot B_{oo} \qquad B_{oo.r.tot.e}(f) := 20 \cdot \log\left(\left|B_{oo.r.tot}(f)\right|\right)$$

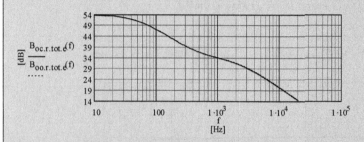

Figure 15.58 Frequency responses of the two pre-amps with
passive RIAA network

$$B_{oc.r.tot.e}(20Hz) = 53.228 \times 10^{0} \qquad\qquad B_{oo.r.tot.e}(20Hz) = 53.229 \times 10^{0}$$

$$B_{oc.r.tot.e}\left(10^{3}Hz\right) = 33.956 \times 10^{0} \qquad\qquad B_{oo.r.tot.e}\left(10^{3}Hz\right) = 34.003 \times 10^{0}$$

$$B_{oc.r.tot.e}\left(20 \cdot 10^{3}Hz\right) = 14.344 \times 10^{0} \qquad\qquad B_{oo.r.tot.e}\left(20 \cdot 10^{3}Hz\right) = 14.453 \times 10^{0}$$

15.5.5 Deviation $D_{xy.r}(f)$ from the exact RIAA transfer $R_0(f)$:

$$B_{oc.r.tot.dif.e}(f) := B_{oc.r.tot.e}(f) - B_{oc.r.tot.e}\left(10^{3}Hz\right)$$

$$B_{oo.r.tot.dif.e}(f) := B_{oo.r.tot.e}(f) - B_{oo.r.tot.e}\left(10^{3}Hz\right)$$

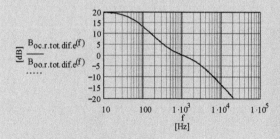

Figure 15.59 RIAA transfers ref. 0dB/1kHz

➢   MCD Worksheet XV-3   local feedback versions (RIAA)   page 12

$$R_{1000} := \frac{\sqrt{1 + \left(2 \cdot \pi \cdot 10^3 Hz \cdot 318 \cdot 10^{-6} s\right)^2}}{\sqrt{1 + \left(2 \cdot \pi \cdot 10^3 Hz \cdot 3180 \cdot 10^{-6} s\right)^2} \sqrt{1 + \left(2 \cdot \pi \cdot 10^3 Hz \cdot 75 \cdot 10^{-6} s\right)^2}}$$

$$R_0(f) := R_{1000}^{-1} \cdot \frac{\sqrt{1 + \left(2 \cdot \pi \cdot f \cdot 318 \cdot 10^{-6} s\right)^2}}{\sqrt{1 + \left(2 \cdot \pi \cdot f \cdot 3180 \cdot 10^{-6} s\right)^2} \sqrt{1 + \left(2 \cdot \pi \cdot f \cdot 75 \cdot 10^{-6} s\right)^2}}$$

$$D_{oc.r}(f) := 20 \cdot \log\left(R_0(f)\right) - B_{oc.r.tot.dif.e}(f) \qquad D_{oo.r}(f) := 20 \cdot \log\left(R_0(f)\right) - B_{oo.r.tot.dif.e}(f)$$

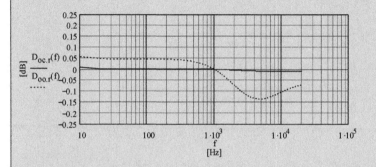

Figure 15.60 Deviations from the exact RIAA transfer

15.6.6 Input impedances $Z_{in.B.xy}(f)$ of the 2 versions:

$$Z_{in.B.oc}(f) := \left[\frac{1}{R1a} + \cfrac{1}{Z1_1(f) + \left(\cfrac{1}{R1b} + \cfrac{1}{R2 + Z2_{1.oc}(f)} \right)^{-1}} \right]^{-1}$$

$$Z_{in.B.oo}(f) := \left[\frac{1}{R1a} + \cfrac{1}{Z1_1(f) + \left(\cfrac{1}{R1b} + \cfrac{1}{R2 + Z2_{1.oo}(f)} \right)^{-1}} \right]^{-1}$$

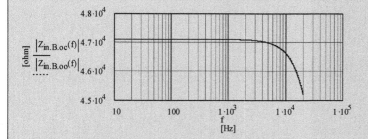

Figure 15.61 Input impedances of the two versions

15.6.7 Final remarks:

15.6.7.1   Varying f_{opt} will lead to drastic changes in the low frequency range of the $B_{oo.r}(f)$ plot - shown in Figure 15.60

15.6.7.2   To compensate the losses of the passive networks between V1 ... 3 gains B_{oc} and B_{oo} must be set > 500. Trimming of R5, 10 will lead to gains of +34dB of $B_{oc.r.tot.e}(f)$ and $B_{oo.r.tot.e}(f)$ - shown in Figure 15.58

15.6.7.3   To get lowest deviation from the exact RIAA transfer only R_A and R_B need trimming - assumed that R_C and C2, C3 got fixed values that were guessed with the respective formulae given in my book "The Sound of Silence"

Chapter 16 Valve data[1] (DS)

Given is a selection of data sheet characteristic charts of valves used in the example calculations. As long as there exist no other sources (like in the solid state world) to simulate valves for calculation purposes these types of charts are the only sources to get the valve constants at certain plate-cathode voltages.

	EU	US	valves with equal data but heater differences		
16.1	E188C[2]	7308	ECC88[3]	6DJ8	
			E88CC[4]	6922	CCa
			PCC88[5]	7DJ8	
			UCC88[6]		
16.2	EC92	6AB4			
16.3	ECC83	12AX7	6AX7		

---

[1]   Data concerning valve related capacitances are not given in this chapter because they can easily be found in valve data hand books like eg. "Röhren Taschentabelle" (valve pocket table), Franzis-Verlag 1963, Munich, Germany, or in the internet: "http://frank.pocnet.net/sheetsE1.html"

[2]   heater voltage and current: 6.3V, 325 mA

[3]   heater voltage and current: 6.3V, 365 mA

[4]   heater voltage and current: 6.3V, 300 mA

[5]   heater voltage and current: 7.0V, 300 mA

[6]   heater voltage and current: 21.0V, 100 mA

16.1   E188CC / 7308 [7]

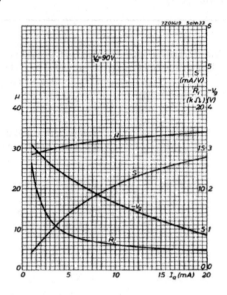

Figure 16.1 E188CC / 7308 valve constants vs. plate current
at 90V plate-cathode voltage

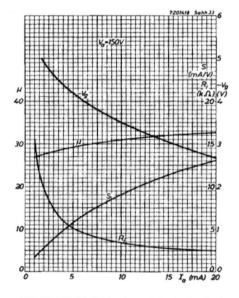

Figure 16.2 E188CC / 7308 valve constants vs. plate current
at 150V plate-cathode voltage

---

[7]   characteristic charts taken from the 1968 Philips Data Handbook on Electronic Components and
Materials; S=g_m, R_i=r_a

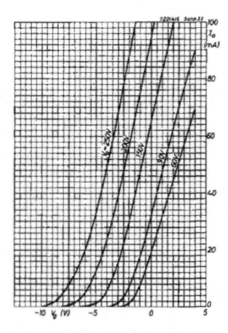

Figure 16.3 E188CC / 7308 grid voltage vs. plate current
and plate-cathode voltage

16.2   EC92 / 6AB4[8]

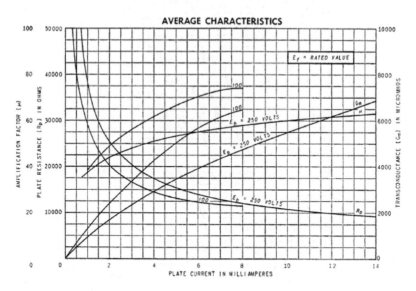

Figure 16.4   EC92 / 6AB4 valve constants vs. plate current
at 250V and 100V plate-cathode voltage

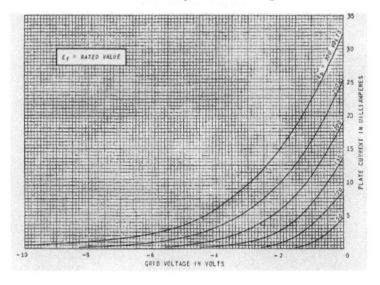

Figure 16.5 EC92 / 6AB4 grid voltage vs. plate current and plate-cathode voltage

---

[8]   characteristic charts taken from the 1956 General Electric data sheet

1632 ECC83 / 12AX7[9]

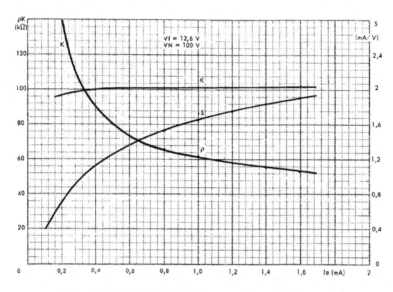

Figure 16.6 ECC83 / 12AX7 valve constants vs. plate current
at 100V plate-cathode voltage

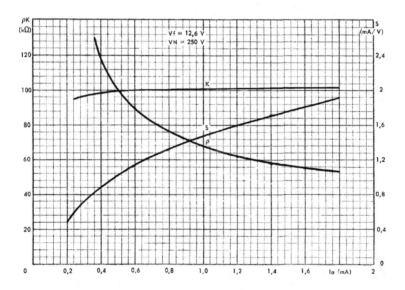

Figure 16.7 ECC83 / 12AX7 valve constants vs. plate current
at 250V plate-cathode voltage

---

[9] characteristic charts taken from the 1968 MAZDA-BELVU data sheet

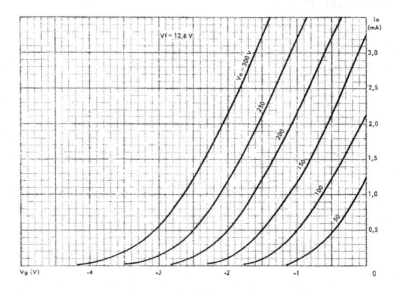

Figure 16.8 ECC83 / 12AX7 grid voltage vs. plate current and plate-cathode voltage

Chapter 17   Book Ending Sections

17.1   Abbreviations:

A	= auxiliary term of a function or equation
A	= example version of Chapter 15
AC	= alternating current, stands also for small signal
AIK	= Aikido gain stage
amp	= amplifier
atan	= arctan in MCD
B	= bandwidth
B	= auxiliary term of a function or equation
B	= example version of Chapter 15
BAL	= balanced gain stage
B_{20k}	= in a bandwidth of 20Hz … 20 kHz = audio band
CFx	= type of cathode follower
C	= capacitance
$C_{g.a}$	= capacitance between grid and plate
$C_{g.c}$	= capacitance between grid and cathode
$C_{a.c}$	= capacitance between plate and cathode
C_i	= input capacitance
C_{in}	= i/p capacitor that keeps DC voltage from the previous stage off the gate
$C_{i.tot}$	= total input capacitance
C_o	= output capacitance
C_{out}	= o/p capacitor that keeps DC voltage off R_L
C_{stray}	= stray capacitances
CAS	= cascode amp stage
CCF	= cascaded cathode follower
CF	= cathode follower
CM	= Miller capacitance
CMRR	= common mode rejection ratio
CCS	= common cathode stage
CCS+Cc	= common cathode stage with cathode AC grounded
CGS	= common grid stage
CSi	= current sink
CSo	= current source
DC	= direct current
deg	= degrees
DS	= data sheet(s)
EX	= example
f	= page plus the following one
(f)	= a frequency dependent function (eg. T(f))
ff	= page plus following ones
$g_{m.t}$	= triode's mutual conductance (t = triode's specific indication)
G	= gain factor of a gain stage
Gen	= Generator
hp	= high-pass

i	= AC current
i_a	= anode or plate AC current
i_c	= cathode AC current
i/p	= input
I	= DC current
Im	= imaginary component of an equation
k	= kilo decimal point of a resistor, eg. 5k62 = 5.62 kΩ
lp	= low-pass
M	= Mega decimal point of a resistor, eg. 5M62 = 5.62 MΩ
M	= auxiliary term of a function or equation
MCD	= MathCad
MM	= Moving Magnet cartridge
n	= nano decimal point of a capacitance, eg. 22n1 = 22.1nF
N	= auxiliary term of a function or equation
o/p	= output
O	= auxiliary term of a function or equation
p	= pico decimal point of a capacitance, eg. 22p1 = 22.1pF
P	= auxiliary term of a function or equation
PAR	= valves in paralleled operation
PEN	= pentode
Q	= auxiliary term of a function or equation
$r_{a.t}$	= plate or anode resistance (= triode's internal resistance)
$r_{c.t}$	= cathode resistance (= triode's internal resistance)
rot	= rule of thumb
$r_{g.t}$	= grid input resistance (= triode's internal resistance)
R	= resistance
R	= decimal point of a resistor, eg. 5R62 = 5.62 Ω
R	= auxiliary term of a function or equation
R_a	= plate loading resistor
R_c	= bias resistance for the cathode = cathode resistor
Re	= real component of an equation
$R_{i.c}$	= cathode input resistance
$R_{o.a}$	= plate output resistance
$R_{o.c}$	= cathode output resistance
R_g	= bias resistance for the grid = grid resistor
R_{gg}	= oscillation prevention resistor at the grid
R_L	= gain stage load resistance = input resistance of next stage
(R_L)	= a R_L dependent function (eg. $G(R_L)$)
R_S	= source impedance
SRPP	= shunt regulated push-pull gain stage
T	= transfer function
TSOS	= The Sound of Silence (ISBN 978-3-540-76883-8)
v	= AC voltage
V	= DC voltage
$V_{a.t}$	= DC voltage between plate and ground (0V)-level
$V_{ax.t}$	= DC voltage between plate and cathode in a multi (x) valve arrangement
$V_{c.t}$	= DC voltage between cathode and ground (0V)-level
$V_{g.t}$	= DC voltage between grid and ground (0V)-level
$V_{gx.t}$	= DC voltage between grid and ground (0V)-level in a multi (x) valve arrangement or between grid and cathode as specifically indicated in the respective illustration
Vx	= valve number x
$v_{a.c}$	= AC voltage between plate and cathode = plate-cathode AC voltage

$v_{g.c}$	= AC voltage between grid and cathode = grid-cathode AC voltage
$v_{R.c}$	= AC voltage across R_c
v_i	= AC input voltage of the gain stage
v_o	= AC output voltage of gain stage
V_{cc}	= positive gain stage DC supply voltage
V_{ee}	= negative gain stage DC supply voltage
V_x	= valve x
v_0	= idle voltage of a source (0=zero)
WCF	= White cathode follower
μ	= micro decimal point of a capacitance, eg. $22\mu 1 = 22.1\mu F$
μ_t	= triode gain (assumption: plate loading resistance R_a is infinite high
μ-F	= μ-follower gain stage
Z	= impedance
$Z_{i.c}$	= cathode input impedance of the gain stage
$Z_{i.g}$	= grid input impedance of the gain stage
$Z_{o.a.eff}$	= effective plate output impedance of gain stage
$Z_{o.c.eff}$	= effective cathode output impedance of gain stage
φ	= phase angle in rad (in MCD)
φ/deg	= phase angle in degrees (in MCD)

17.2   Subscripts:

a	= anode = plate
a.c	= anode/plate-cathode ….
aik	= Aikido gain stage
app	= approximation, approximated
b	= bypassed
b	= balanced operation
c	= cathode
cas	= cascode amp
c.t	= cathode of a specific triode t
cc	= positive
cc	= overall feedback plus local feedback around each valve
ccf	= cascoded cathode follower
cf	= cathode follower
cf1	= cf simple version
cf2	= cf improved version
co	= overall feedback plus no local feedback for V2 only
e	= logarithmic expression (e.g. $H_e = 20\log(H)$)
ee	= negative
eff	= effective
g	= for grid bias purposes
gg	= oscillation prevention component via grid
g.a	= grid-plate ….
g.c	= grid-cathode ….
gx	= grid of valve x in a multi valve (x = 1 … n)) arrangement
g1	= grid 1 of pentode
g2	= grid 2 of pentode
g3	= grid 3 of pentode
hi	= high-Z (current source or sink)
i	= in
in	= in
i.t	= i/p of a specific triode t
lo	= low-Z (current source or sink)
L	= load
m	= mutual
n	= new
n	= n times
o	= out
o.a	= plate or anode o/p
oc	= no overall feedback plus local feedback for V1 and V2
o.c	= cathode o/p
oo	= no overall feedback plus no local feedback for V1 and V2
opt	= optimal
o.t	= o/p of a specific triode t
out	= out
p	= peak
p	= pentode
par	= parallel operation
r	= RIAA equalized
rot	= rule of thumb

R	= resistor
red	= reduce, reduced
srpp	= SRPP gain stage
stray	= stray capacitance
S	= source
t	= stands for triode
tot	= total
u	= un-bypassed
u	= unbalanced operation
$\mu.F$	= μ-Follower gain stage
vx	= valve number x (Vx)
wcf	= White cathode follower
0	= source voltage number 0
1	= one (times)
2	= two (times)
20k	= 20Hz … 20kHz
83	= short form identification for ECC83 / 12AX7 in calculations
92	= short form identification for EC92 / 6AB4 in calculations
188	= short form identification for E188CC / 7308 in calculations
.	= in Figures the first point indicates a subscript, following points separate different aspect namings, in all other cases any point separates different aspect namings

17.3   **List of Figures:**
(Figures on text pages and *on MCD worksheets*)

17.4   Index

Note: A page number followed by 'f' or 'ff' indicates that the topic is also discussed in the following page(s).

17.5   Epilogue

The whole thing started after I've finished writing my book[1] on noise in audio pre-amps in September 2007. I tried to calculate the noise of a moving magnet RIAA gain stage that is based on the SRPP concept, developed 1969 by Mr Anzai from Japan. His recommendation to take a ECC81 / 12AT7 as the first stage valve was not followed by Elektor Electronics in their 1993 book on high-end audio[2]. Their proposed solution for the first valve was a ECC83 / 12AX7. This triggered my intention to find out the difference - especially on noise. To calculate the noise of these two-stage SRPP designs we need to know the gains of the stages as well as the loss of the RIAA networks.

I discovered that the gain formulae I could find in many sources (books, internet, magazines, etc. - my own university papers got teared to pieces long time ago) where not satisfying enough to calculate things with the precision I wanted to get. Most descriptions were too complicate. Or they were rather distorted and hidden in a lot of hobby blah-blah. Or they offered too many simplifications. In addition, many formulae where simply wrong and I couldn't find a source that offers the broad range of triode gain stage possibilities on one spot - in a manner I was dreaming of. No source gave easy to handle formulae to calculate transfer functions - from input source to output load - of complete gain stages covering only one or several valves in a sequence.

Finally, for my own purposes, I decided to start to develop all formulae that are capable to describe the behaviour of many different triode gain stages in a way that makes calculation life for me - and later on for the audio enthusiast, developer and / or - not only young - engineer much more easy - with the help of a mathematical software and the respective worksheets, thus, virtually producing a kind of simulation approach. And, most important of all my efforts, one should be able to follow the derivation course. Despite the many subscripts and abbreviations one should get a better understanding of the whole matter. The credo I was following up became: "calculate first and do the roundings not before the end".

But I must point out that the calculation examples I'm presenting on MathCad worksheets are examples to show one possible calculation path only! They do not represent the top result of all development efforts one could imagine. It will be the task of the user to find out the best results by eg. changing the circuit variables and bias data - on a strictly own development basis.

Coming back to the results of my comparative SRPP calculations: from a noise point of view the Signal-to-Noise-Ratio (SN) of the ECC81 / 12AT7 1st stage approach is approximately 2dB better in a 20Hz ... 20kHz frequency band than the ECC83 / 12AX7 solution. It was calculated with a 1k input load and with A-weighting. A second calculation with a Shure V15 V MM cartridge as the input load ended up with nearly no difference in SN, hence, for low impedance cartridges the ECC81 / 12AT7 input device is the better solution. This is also valid for input

---

[1]   TSOS (see Section 17.1)
[2]   "High-end Audio Equipment", Elektor Electronics UK, 1993,
          ISBN 0-905-70540-8

step-up transformers for MC cartridge puposes. The price to pay is the nearly doubled 1st stage DC current, meanwhile the overall gain doesn't change significantly.

At the end of this reference book I wish all readers a lot of fun with triodes - despite the nasty math! Old-timers never die! They reflect a certain emotional value and for many connoisseurs the sound of eg. a 1976 "911" triggers the equivalent emotions like the soft glow of valves in the background of a living room and the right music that enchant ones ears.

I guess that's what many valve enthusiasts are living for!